Wan Fokkink

Introduction
to Process Algebra

With 11 Figures and 11 Tables

Springer
Berlin
Heidelberg
New York
Barcelona
Hong Kong
London
Milan
Paris
Singapore
Tokyo

Author

Dr. Wan Fokkink

Centrum voor Wiskunde en Informatica (CWI)
P.O. Box 94079
1090 GB Amsterdam, The Netherlands
wan@cwi.nl
http://www.cwi.nl/~wan

Library of Congress Cataloging-in-Publication Data applied for

Die Deutsche Bibliothek - CIP-Einheitsaufnahme

Fokkink, Wan:
Introduction to process algebra/Wan Fokkink. - Berlin; Heidelberg;
New York; Barcelona; Hong Kong; London; Milan; Paris;
Singapore; Tokyo: Springer 2000
    (Texts in theoretical computer science)

ACM Computing Classification (1998): D.2.4, D.3.1–3, F.3.2, F.1.2, F.4.2

ISBN 978-3-642-08584-0

© Springer-Verlag Berlin Heidelberg 2010
Printed in Germany

Cover Design: Künkel + Lopka, Werbeagentur, Heidelberg

Preface

Computer software and network protocols are increasingly important in daily life. At the same time the complexity of software has rocketed, so that its correctness is at stake. New methodologies and disciplines are being developed to bring structure to the ever growing jungle of computer technology. (Semi-)automated manipulation has become an important means in discovering flaws in software and hardware systems. Process algebra is a mathematical framework in which system behaviour is expressed in the form of algebraic terms, enhancing the available techniques for manipulation.

Concurrency is omnipresent in system behaviour, and in a large part responsible for its complexity: even simple behaviours become wildly complicated when they are executed in parallel. In order to study such systems in detail, it is imperative that they are dissected into their concurrent components. Fundamental to process algebra is a parallel operator, to break down systems into their concurrent components. A set of equations is imposed to derive whether two terms are behaviourally equivalent. In this framework, non-trivial properties of systems can be established in an elegant fashion. For example, it may be possible to equate an implementation to the specification of its required input/output relation. In recent years a variety of automated tools have been developed to facilitate the derivation of such properties.

Applications of process algebra exist in diverse fields such as safety critical systems, network protocols, and biology. In the educational vein, process algebra has been recognised to teach skills to deal with complex concurrent systems, by representing and reasoning about such systems in a mathematically clear and precise manner.

This text developed from an undergraduate course on process algebra at the computer science department of the University of Wales Swansea during the autumn of 1997 and of 1998. Chapters 2-7 contain sufficient material for more than twenty hours of lecturing; a set of slides and further material to support such a course are available from my homepage (currently at http://www.cwi.nl/~wan). It is recommended to use a tool set based on process algebra, such as the μCRL tool set or the Concurrency Workbench Edinburgh, to enliven the course. Appendices A and B provide useful background information; they are not intended to be included in the course.

I am grateful to John Tucker for his encouragement to further develop a raw set of lecture notes, and to Judi Romijn for her support. Over the years I have benefited from discussions with Jan Bergstra, Rob van Glabbeek, Jan Friso Groote, Frits Vaandrager, Alban Ponse, Chris Verhoef, Jaco van de Pol, Jos Baeten, Luca Aceto, Jos van Wamel, Steven Klusener, Bas Luttik, Dennis Dams, and many others.

Amsterdam, November 1999   *Wan Fokkink*

Contents

1. Introduction

System behaviour generally consists of processes and data. Processes are the control mechanisms for the manipulation of data. While processes are dynamic and active, data are static and passive. System behaviour tends to be composed of several processes that are executed concurrently, where these processes exchange data in order to influence each other's behaviour. The picture below presents a typical architecture for a concurrent system. Each process P_i sends messages to its neighbouring processes P_{i-1} and P_{i+1}, giving them information on the state of P_i. The neighbouring processes use this information in their internal computations, to update their own states.

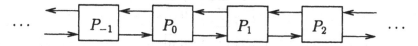

Some examples of concurrent systems are:

- A colony of ants:

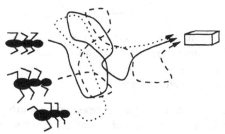

  Ants behave as separate entities, which influence each other's behaviour. As soon as one ant discovers a goody such as a lump of sugar, it radiates a smell to attract other ants. Tofts [186] was able to explain certain phenomena of colonies of ants by modelling such colonies as concurrent systems in process algebra.
- A network protocol, being a high-level description of a data communication procedure.

  As an example we consider the so-called alternating bit protocol [31]. A Sender and a Receiver are the separate processes, which in concurrency make up the system:

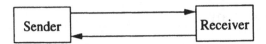

Data elements $d_1, d_2, d_3, \ldots$ are sent from the Sender to the Receiver via a faulty channel, so that data may be corrupted. In the alternating bit protocol, the Sender attaches a bit 0 to data elements d_{2k-1} and a bit 1 to data elements d_{2k} for positive natural numbers k. As soon as the Receiver receives a datum, it sends the attached bit to the Sender via a faulty channel, to acknowledge reception. If the Receiver receives a corrupted message, then it resends the previous acknowledgement. The Sender keeps on sending out the pair (d_i, b) until it receives the acknowledgement b. Then it starts sending out the next pair $(d_{i+1}, 1 - b)$ until it receives the acknowledgement $1 - b$, et cetera. Alternation of the attached bit enables the Receiver to determine whether a received datum is really new, and alternation of the acknowledgement enables the Sender to determine whether a datum reached the Receiver unscathed.

- A pocket calculator:

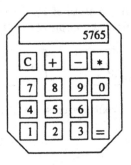

The buttons represent the separate actions of this system, which all influence the state (i.e., the intermediate result of a computation) of the pocket calculator in a different way. The pocket calculator in combination with a user make up a concurrent system.

In this text, system behaviour is represented as a labelled transition system, which basically consists a set of nodes together with a set of labelled edges between these nodes. For example, a fraction of the full labelled transition system of the pocket calculator is depicted in Fig. 1.1. Each node in this labelled transition system represents a different state of the calculator, and an edge from one node to the other expresses that the calculator can change from one state to the other, by pushing a button; the label of an edge represents the button that has to be pushed in order to realise this state transition.

In general it is much easier to study a concurrent system such as the pocket calculator by breaking it up into its concurrent components. Although its full labelled transition system is enormous, the process behaviour of the

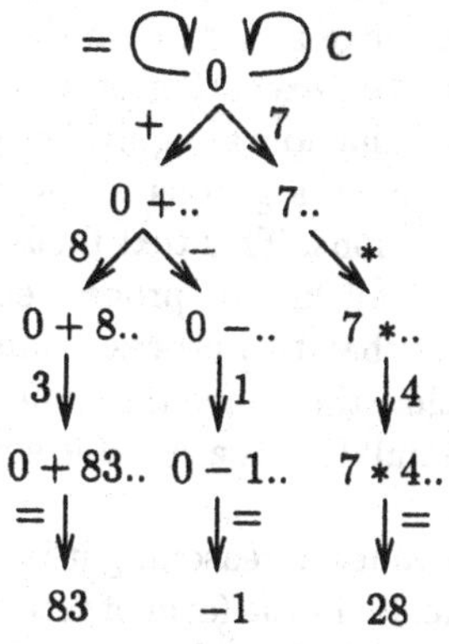

$$= \quad 0 \quad C$$

$$+ \swarrow \searchover 7$$

$$0 +.. \qquad 7..$$

$$8 \swarrow \quad - \searrow *$$

$$0 + 8.. \quad 0 -.. \quad 7 *..$$

$$3 \downarrow \qquad \downarrow 1 \qquad \downarrow 4$$

$$0 + 83.. \quad 0 - 1.. \quad 7 * 4..$$

$$= \downarrow \qquad \downarrow = \qquad \downarrow =$$

$$83 \qquad -1 \qquad 28$$

Fig. 1.1. Labelled transition system of a pocket calculator

pocket calculator is not so difficult. It can be captured by specifying the behaviour of the separate buttons, and putting them in parallel. For example, the behaviour of the +-button is displayed in Fig. 1.2, where $d_1, \ldots, d_k$ are digits and $m = n + d_1 \cdots d_k$. Execution is started in the state that is pictured at the top, where the computation has the intermediate value n. Similarly, the arithmetic operations subtraction and multiplication can be specified on the data domain of numbers. An extra error element needs to be added to the data domain, to represent that the result of an arithmetic computation exceeds the screen size, or that an operation is undefined (such as division by zero).

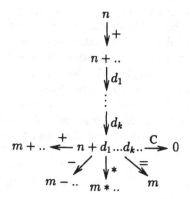

Fig. 1.2. Behaviour of the plus button

A process graph is a labelled transition system in which one state is selected to be the root state, i.e., the initial state of the process. If the labelled transition system contains an edge $s \xrightarrow{a} s'$, then the process graph can evolve from state s into state s' by the execution of action a. Process graphs are distinguished modulo some behavioural equivalence. For example, such an equivalence may relate two process graphs if and only if they can execute exactly the same strings of actions. This text focuses on bisimulation equivalence, which is the finest of all known process equivalences. Bisimulation equivalence requires not only that two process graphs can execute the same strings of actions, but also that they have the same branching structure. Experience has shown that bisimulation is a suitable equivalence when reasoning about concurrent processes.

For the purpose of mathematical reasoning it is often convenient to represent process graphs algebraically in the form of terms. Process algebra focuses on the specification and manipulation of process terms as induced by a collection of operator symbols. This symbolic notation facilitates manipulation by a computer. Most process algebras contain basic operators to build finite processes, communication operators to express concurrency, and some notion of recursion to capture infinite behaviour. Moreover, it is convenient to introduce two special constants: the deadlock enables us to force actions into communication, while the silent step allows us to abstract away from internal computations. Structural operational semantics is used to formally provide each process term over these operators and constants with its intended process graph. The crux of process algebra is that it imposes an equational logic on process terms, such that two process terms can be equated if and only if their graphs are behaviourally equivalent. A process algebra can be extended with fresh operators, to enhance its expressivity or to facilitate the specification of system behaviour. Such a fresh operator requires an extension of the structural operational semantics and of the equational logic.

Process algebra constitutes a framework for formal reasoning about processes and data, with the emphasis on processes that are executed concurrently. It can be used to detect undesirable properties and to formally derive desirable properties of a system specification. Notably, process algebra can be used to verify that a system displays the desired external behaviour, meaning that for each input the correct output is produced. First, the implementation of the system is expressed in the form of a process term, using the basic operators, the communication operators, and recursion. Next, the deadlock is used to force actions into communication, and the silent step is used to abstract away from internal computations, so that only the input/output relation of the implementation remains. Finally, the resulting process term is manipulated by means of equational logic, to prove that its graph conforms with the desired external behaviour.

The foundations of process algebra were developed, largely independently, by Milner [149, 150, 151] and Hoare [125, 126, 127]. These foundations are partly rooted in Petri nets [168], automata theory [178], formal languages [7], and work by Bekič [33]. Milner devised the process algebra CCS (Calculus of

Communicating Systems) [155], while Hoare pioneered CSP (Communicating Sequential Processes) [177]. The current exposition is based on the approach of Bergstra and Klop [41] called ACP (Algebra of Communicating Processes) [28], which is closely related to CCS. Interesting early accounts of ACP are [47, 48, 50].

Data and time often play an important role in system behaviour. Similar to processes, data can be specified algebraically by means of an equational logic; see [51, 143]. In this text it is usually assumed implicitly that the data types have been specified beforehand. Furthermore, in an example verification, time is modelled using special timer processes, which can pass on timing information. Alternatively, time could be modelled by adding time stamps to actions, to fix the moment in time at which such an action can be executed, and adapting the semantics to take into account such timing information.

Some expositions on process algebra, notably the one by Baeten and Weijland [10, 28], start by defining a set of equations, and give semantic models for which this equational logic is sound and complete, meaning that two process terms can be equated if and only if they are equivalent in the model. Advantages of this approach are that results for several models can be derived simultaneously, and the emphasis that process algebra is relatively independent of its models. Following for instance Milner [155] and Baeten and Verhoef [27], the focus in this text is on a single model, based on structural operational semantics. An advantage of the latter approach is that it allows us to place the exposition more firmly on an intuitive basis.

This text is set up as follows. Chapter 2 introduces basic process algebra, which can express finite process graphs; an equational logic is presented that is sound and complete modulo bisimulation equivalence. Chapter 3 features merge operators to express processes that are executed in parallel. Chapter 4 defines recursion to describe infinite process behaviour. Chapter 5 explains how to abstract away from internal computations. Chapter 6 applies the framework from the previous chapters to verify the correctness of two network protocols. Furthermore, it gives an overview of existing techniques and automated tools to support such verification efforts. Chapter 7 gives examples of further operators that can be added to the framework.

Appendices A and B provide background material for the theory developed in the earlier chapters. Appendix A presents the basics of equational logic, while Appendix B gives an overview of structural operational semantics. The reader is adviced to use these appendices to become acquainted with basic notions and definitions when they are encountered in the remaining chapters. Pointers to relevant definitions in the appendices are given when appropriate.

2. Basic Process Algebra

This chapter presents a basic framework for process algebra. It introduces simple operators that enable us to construct finite processes from scratch.

2.1 Basic Process Terms

The signature (see Definition A.1.1) of a basic framework for process algebra consists of the following operators.

- First of all, we assume a finite, non-empty set A of *(atomic) actions*, representing indivisible behaviour (such as reading a datum, or sending a datum). Each atomic action a is a constant that can execute itself, after which it terminates successfully:

  The predicate $\overset{a}{\to} \sqrt{}$ represents successful termination after the execution of action a.

- Moreover, we assume a binary operator $+$ called *alternative composition*. If closed terms t_1 and t_2 (see Definition A.1.2) represent processes p_1 and p_2 (see Definitions B.1.1 and B.3.1), respectively, then the closed term $t_1 + t_2$ represents the process that executes either p_1 or p_2. In other words, the process graph of $t_1 + t_2$ is obtained by joining p_1 and p_2 at their root states:

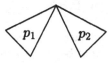

- Finally, we assume a binary operator $\cdot$ called *sequential composition*. If closed terms t_1 and t_2 represent processes p_1 and p_2, respectively, then the closed term $t_1 \cdot t_2$ represents the process that executes first p_1 and then p_2. In other words, the process graph of $t_1 \cdot t_2$ is obtained by replacing each successful termination $s \overset{a}{\to} \sqrt{}$ in p_1 by a transition $s \overset{a}{\to} s'$, where s' is the root of p_2:

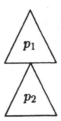

Example 2.1.1. Let a, b, c, and d be actions. The closed term $((a + b) \cdot c) \cdot d$ represents the following process, with the root state presented at the top:

Each finite process (see Definition B.3.1) can be represented by a closed term that is built from the set A of atomic actions, the $+$, and the $\cdot$. Such terms are called *basic process terms*, and the collection of all basic process terms is called *basic process algebra*, abbreviated to BPA.

Exercise 2.1.1. Find the basic process terms that belong to the following two process graphs (with their root states presented at the top):

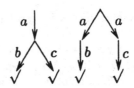

2.2 Transition Rules for BPA

We have provided a syntax for basic process terms, together with some intuition for the process graph that belongs to such a term. This relationship has to be made formal in order for it to become really meaningful. For this purpose we apply *structural operational semantics*, as explained in Appendix B. This involves giving a collection of transition rules (see Definition B.1.2), which define transitions $t \xrightarrow{a} t'$ (see Definition B.1.1) to express that term t can evolve into term t' by the execution of action a, and predicates $t \xrightarrow{a} \sqrt{}$ to express that term t can terminate successfully by the execution of action a.

Table 2.1 presents the TSS (see Definition B.1.2) that constitutes the structural operational semantics of BPA. The variables x, x', y, and y' in the transition rules range over the collection of basic process terms, while v ranges over the set A of atomic actions.

Table 2.1. Transition rules of BPA

$$v \xrightarrow{v} \surd$$

$$\frac{x \xrightarrow{v} \surd}{x + y \xrightarrow{v} \surd} \quad \frac{x \xrightarrow{v} x'}{x + y \xrightarrow{v} x'} \quad \frac{y \xrightarrow{v} \surd}{x + y \xrightarrow{v} \surd} \quad \frac{y \xrightarrow{v} y'}{x + y \xrightarrow{v} y'}$$

$$\frac{x \xrightarrow{v} \surd}{x \cdot y \xrightarrow{v} y} \quad \frac{x \xrightarrow{v} x'}{x \cdot y \xrightarrow{v} x' \cdot y}$$

The TSS of BPA provides each basic process term with a process graph, according to the intuition that was presented in the previous section:

- the first transition rule says that each atomic action v can terminate successfully by executing itself;
- the next four transition rules express that $t + t'$ executes either t or t';
- the last two transition rules express that $t \cdot t'$ executes t until successful termination, after which it proceeds to execute t'.

Example 2.2.1. The transition rules in Table 2.1 provide the basic process term $((a + b) \cdot c) \cdot d$ with the following process graph (cf. Example 2.1.1):

$$((a + b) \cdot c) \cdot d$$

For instance, the transition $((a + b) \cdot c) \cdot d \xrightarrow{b} c \cdot d$ can be proved (see Definition B.1.3) from the transition rules in Table 2.1 as follows:

$$b \xrightarrow{b} \checkmark \qquad (\frac{}{v \xrightarrow{v} \checkmark}, \qquad v := b)$$

$$\frac{}{a + b \xrightarrow{b} \checkmark} \qquad (\frac{y \xrightarrow{v} \checkmark}{x + y \xrightarrow{v} \checkmark}, \qquad v := b, \ x := a, \ y := b)$$

$$\frac{}{(a + b) \cdot c \xrightarrow{b} c} \qquad (\frac{x \xrightarrow{v} \checkmark}{x \cdot y \xrightarrow{v} y}, \qquad v := b, \ x := a + b, \ y := c)$$

$$\frac{}{((a + b) \cdot c) \cdot d \xrightarrow{b} c \cdot d} \ (\frac{x \xrightarrow{v} x'}{x \cdot y \xrightarrow{v} x' \cdot y}, v := b, \ x := (a + b) \cdot c,$$
$$x' := c, \ y := d)$$

At the right-hand side, the transition rules are displayed that are applied in the consecutive proof steps, together with the closed substitutions (see Definition A.1.3) that are applied to them.

Exercise 2.2.1. Find the process graph that belongs to the basic process term $((a + b) \cdot (a + c)) \cdot d$. Give the derivations of the transitions in this process graph from the transition rules in Table 2.1.

From now on, as binding convention we assume that the $\cdot$ binds stronger than the $+$. For example, $a \cdot b + a \cdot c$ represents $(a \cdot b) + (a \cdot c)$. Occurrences of $\cdot$ are often omitted from process terms; that is, st denotes $s \cdot t$.

2.3 Bisimulation Equivalence

In the previous section, each basic process term has been provided with a process graph using structural operational semantics. Processes have been studied since the early 60's, first to settle questions in natural languages, later on to study the semantics of programming languages. These studies were in general based on so-called trace equivalence, in which two processes are said to be equivalent if they can execute exactly the same strings of actions. However, for system behaviour this equivalence is not always satisfactory, which is shown by the following example.

Example 2.3.1. Consider the two processes below:

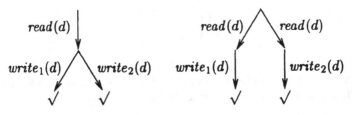

The first process reads datum d, and then decides whether it writes d on disc 1 or on disc 2. The second process makes a choice for disc 1 or disc 2 before it reads datum d. Both processes display the same strings of actions, $read(d)write_1(d)$ and $read(d)write_2(d)$, so they are trace equivalent. Still, there is a crucial distinction between the two processes, which becomes apparent if for instance disc 1 crashes. In this case the first process always saves datum d on disc 2, while the second process may get into a deadlock (i.e., may get stuck).

Bisimulation equivalence (see Definition B.3.2) discriminates more processes than trace equivalence. Namely, if two processes are *bisimilar*, then not only they can execute exactly the same strings of actions, but also they have the same branching structure. For example, the two processes in Example 2.3.1 are not bisimilar. Definition B.3.2 is presented below for the relations $\overset{a}{\rightarrow}$ and the predicates $\overset{a}{\rightarrow} \checkmark$, for $a \in A$.

Definition 2.3.1 (Bisimulation). *A* bisimulation relation $\mathcal{B}$ *is a binary relation on processes such that:*

1. *if $p \mathcal{B} q$ and $p \overset{a}{\rightarrow} p'$, then $q \overset{a}{\rightarrow} q'$ with $p' \mathcal{B} q'$;*
2. *if $p \mathcal{B} q$ and $q \overset{a}{\rightarrow} q'$, then $p \overset{a}{\rightarrow} p'$ with $p' \mathcal{B} q'$;*
3. *if $p \mathcal{B} q$ and $p \overset{a}{\rightarrow} \checkmark$, then $q \overset{a}{\rightarrow} \checkmark$;*
4. *if $p \mathcal{B} q$ and $q \overset{a}{\rightarrow} \checkmark$, then $p \overset{a}{\rightarrow} \checkmark$.*

Two processes p and q are bisimilar, *denoted by $p \leftrightarrow q$, if there is a bisimulation relation $\mathcal{B}$ such that $p \mathcal{B} q$.*

Example 2.3.2. $(a + a)b \leftrightarrow ab + a(b + b)$.
A bisimulation relation that relates these two basic process terms is defined by $(a + a)b \mathcal{B} ab + a(b + b)$, $b \mathcal{B} b$, and $b \mathcal{B} b + b$. This bisimulation relation can be depicted as follows:

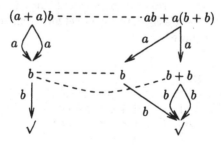

Exercise 2.3.1. Say for each of the following pairs of basic process terms whether they are bisimilar:

- $(b + c)a + ba + ca$ and $ba + ca$;
- $a(b + c) + ab + ac$ and $ab + ac$;
- $(a + a)(bc) + (ab)(c + c)$ and $(a(b + b))(c + c)$.

For each pair of bisimilar terms, give a bisimulation relation that relates them.

Exercise 2.3.2. Show that the basic process terms $read(d) \cdot (write_1(d) + write_2(d))$ and $read(d) \cdot write_1(d) + read(d) \cdot write_2(d)$ are not bisimilar.

Exercise 2.3.3. Prove that $a^{k+1} \not\leftrightarrow a^{k+2}$ for natural numbers k.

Exercise 2.3.4. Verify that bisimilarity is an equivalence relation.

2.4 Axioms for BPA

Checking whether the process graphs of two basic process terms are bisimilar requires hard labour. First these process graphs have to be computed, and next a bisimulation relation has to be established. This section introduces an *axiomatisation* for BPA, to equate bisimilar basic process terms. This avoids the computation of process graphs and bisimulation relations altogether. The axioms have the additional advantage that they can be used in automated reasoning, so that they facilitate a mechanised derivation that two basic process terms are bisimilar.

We are after an axiomatisation (see Definition A.2.1) such that the induced equality relation = (see Definition A.2.2) on basic process terms characterises bisimulation equivalence over BPA in the following sense (cf. Definition A.3.1):

1. the equality relation is *sound*, meaning that if $s = t$ holds for basic process terms s and t, then $s \leftrightarrow t$;
2. the equality relation is *complete*, meaning that if $s \leftrightarrow t$ holds for basic process terms s and t, then $s = t$.

Soundness ensures that if terms can be equated, then they are in the same bisimulation equivalence class, while completeness ensures that bisimilar terms can always be equated.

Table 2.2. Axioms for BPA

A1	$x + y = y + x$
A2	$(x + y) + z = x + (y + z)$
A3	$x + x = x$
A4	$(x + y) \cdot z = x \cdot z + y \cdot z$
A5	$(x \cdot y) \cdot z = x \cdot (y \cdot z)$

Table 2.2 presents an axiomatisation $\mathcal{E}_{BPA}$ for BPA modulo bisimulation equivalence. The variables x, y, and z in the axioms range over the collection

of basic process terms. The equality relation on basic process terms induced by the axiomatisation $\mathcal{E}_{\text{BPA}}$ is obtained by taking the set of closed substitution instances (see Definition A.1.3) of axioms in $\mathcal{E}_{\text{BPA}}$, and closing it under equivalence and contexts; see Definition A.2.2.

Exercise 2.4.1. Prove that the axioms A1-3 are equivalent to axiom A3 together with

$$A2' \quad (x+y)+z = y+(z+x).$$

The equality relation that $\mathcal{E}_{\text{BPA}}$ induces on BPA is closed under contexts. So in order to conclude that this equality relation is sound and complete for BPA modulo bisimulation, we need to know that this equivalence is a *congruence* (see Definition B.3.3) with respect to BPA. That is, if $s \leftrightarrow s'$ and $t \leftrightarrow t'$, then $s+t \leftrightarrow s'+t'$ and $s \cdot t \leftrightarrow s' \cdot t'$.

Theorem 2.4.1. *Bisimulation equivalence is a congruence with respect to BPA.*

Proof. The transition rules in Table 2.1 are in panth format (see Definition B.3.4). So the bisimulation equivalence that they induce is a congruence; see Theorem B.3.1. □

Exercise 2.4.2. Verify that the TSS of BPA is in panth format.

Theorem 2.4.2. $\mathcal{E}_{\text{BPA}}$ *is sound for BPA modulo bisimulation equivalence.*

Proof. Since bisimulation is both an equivalence and a congruence for BPA, we only need to check that the first clause in the definition of the relation $=$ is sound. That is, if $s = t$ is an axiom in $\mathcal{E}_{\text{BPA}}$ and σ a closed substitution that maps the variables in s and t to basic process terms, then we need to check that $\sigma(s) \leftrightarrow \sigma(t)$. We only provide some intuition for soundness of the axioms in Table 2.2:

- A1 (commutativity of $+$) says that both $s+t$ and $t+s$ represent a choice between s and t;
- A2 (associativity of $+$) says that both $(s+t)+u$ and $s+(t+u)$ represent a choice between s, t, and u;
- A3 (idempotency of $+$) says that a choice between t and t amounts to a choice for t;
- A4 (right distributivity of $\cdot$) says that both $(s+t) \cdot u$ and $s \cdot u + t \cdot u$ represent a choice between s and t, followed by u;
- A5 (associativity of $\cdot$) says that both $(s \cdot t) \cdot u$ and $s \cdot (t \cdot u)$ represent s followed by t followed by u.

These intuitions can be made rigorous by means of explicit bisimulation relations between the left- and right-hand sides of closed instantiations of the axioms in $\mathcal{E}_{BPA}$. Hence, all such instantiations are sound modulo bisimulation equivalence. $\square$

Note that left distributivity of sequential composition, i.e., $x \cdot (y + z) = x \cdot y + x \cdot z$, is in general not sound modulo bisimulation equivalence; see Exercise 2.3.2.

Exercise 2.4.3. Prove that $s + t \leftrightarrow t + s$, $(s + t)u \leftrightarrow su + tu$, and $(st)u \leftrightarrow s(tu)$ for all basic process terms s, t, and u.

It remains to prove that $\mathcal{E}_{BPA}$ is complete for BPA modulo bisimulation equivalence, meaning that $s \leftrightarrow t$ implies $s = t$. The following completeness proof is based on turning the axiomatisation $\mathcal{E}_{BPA}$ into a TRS (see Definition A.4.1), by directing the axioms from left to right.

Theorem 2.4.3. $\mathcal{E}_{BPA}$ *is complete for BPA modulo bisimulation equivalence.*

Proof. We consider basic process terms modulo associativity and commutativity (AC) of the $+$, and this equivalence relation is denoted by $=_{AC}$; see Section A.4. That is, $s =_{AC} t$ if and only if s and t can be equated by axioms A1 and A2. A basic process term s then represents the collection of basic process terms t such that $s =_{AC} t$. Each equivalence class s modulo AC of the $+$ can be represented in the form $s_1 + \cdots + s_k$ with each s_i either an atomic action or of the form $t_1 \cdot t_2$; we refer to the subterms $s_1, \ldots, s_k$ as the *summands* of s.

The three remaining axioms A3-5 are turned into rewrite rules, by directing them from left to right:

$$x + x \to x$$
$$(x + y) \cdot z \to x \cdot z + y \cdot z$$
$$(x \cdot y) \cdot z \to x \cdot (y \cdot z)$$

These rewrite rules are applied to basic process terms modulo AC of the $+$. For example, none of the three rewrite rules applies to $(a + b) + a$, but

$$(a + b) + a \ =_{AC} \ b + (a + a) \ \to \ b + a.$$

The TRS is terminating (see Definition A.4.4) modulo AC of the $+$. That is, each reduction of a basic process term ends up in a normal form n (see Definition A.4.3), meaning that the rewrite rules do not apply to any of the basic process terms that are equivalent to n modulo AC of the $+$. This follows from the weight function (cf. Example A.4.2) on basic process terms that is defined inductively as follows, where v ranges over A. The symbol $\stackrel{\Delta}{=}$ stands for "equals by definition".

$$weight(v) \stackrel{\Delta}{=} 2$$
$$weight(s + t) \stackrel{\Delta}{=} weight(s) + weight(t)$$
$$weight(s \cdot t) \stackrel{\Delta}{=} weight(s)^2 \cdot weight(t).$$

(Here, $+$ and $\cdot$ refer to addition and multiplication on the natural numbers, respectively.) It is not hard to see that each application of a rewrite rule strictly decreases the weight of a basic process term, and that moreover basic process terms that are equivalent modulo AC of the $+$ have the same weight. Since each sequence of natural numbers $k_1 > k_2 > k_3 > \cdots$ is finite, it follows that the TRS is terminating modulo AC of the $+$.

Owing to the forms of the left-hand sides of the three rewrite rules, normal forms are built from distinct summands a and as, with a an atomic action and s a normal form. We prove for normal forms n and n' that $n \leftrightarrow n'$ implies $n =_{\text{AC}} n'$. The proof is based on induction with respect to the sizes of n and n', meaning the number of function symbols that they contain. Let $n \leftrightarrow n'$.

- Consider a summand a of n. Then $n \overset{a}{\to} \sqrt{}$, so $n \leftrightarrow n'$ implies $n' \overset{a}{\to} \sqrt{}$, meaning that n' also contains the summand a.
- Consider a summand as of n. Then $n \overset{a}{\to} s$, so $n \leftrightarrow n'$ implies $n' \overset{a}{\to} t$ with $s \leftrightarrow t$, meaning that n' contains a summand at. Since s and t are normal forms and have sizes smaller than n and n', respectively, by induction $s \leftrightarrow t$ implies $s =_{\text{AC}} t$.

Hence, each summand of n is also a summand of n'. Vice versa, each summand of n' is also a summand of n. In other words, $n =_{\text{AC}} n'$.

Finally, let the basic process terms s and t be bisimilar. The TRS is terminating modulo AC of the $+$, so it reduces s and t to normal forms n and n', respectively. Since the rewrite rules and equivalence modulo AC of the $+$ can be derived from the axioms, $s = n$ and $t = n'$. Soundness of the axioms then yields $s \leftrightarrow n$ and $t \leftrightarrow n'$, so $n \leftrightarrow s \leftrightarrow t \leftrightarrow n'$. We showed that $n \leftrightarrow n'$ implies $n =_{\text{AC}} n'$. Hence, $s = n =_{\text{AC}} n' = t$. □

The proof of Theorem 2.4.3 points out a mechanised way to verify whether two basic process terms are bisimilar. First, reduce both basic process terms to a normal form, by means of the rewrite rules. Next, check whether the two resulting normal forms are equivalent modulo AC of the $+$. If so, then the original terms are bisimilar; if not, then the original terms are not bisimilar.

Exercise 2.4.4. Verify for the TRS in the proof of Theorem 2.4.3 that if $s \to t$ then $weight(s) > weight(t)$, and if $s =_{\text{AC}} t$ then $weight(s) \equiv weight(t)$.

Exercise 2.4.5. Suppose the definition of the weight function in the proof of Theorem 2.4.3 would be adapted by putting $weight(s \cdot t) \overset{\Delta}{=} weight(s) \cdot weight(t)$. Give basic process terms s and t of the same weight such that $s \to t$.

Example 2.4.1. We equate the bisimilar basic process terms $(a + b) + a$ and $(b + a) + b$. First, they are reduced to normal form:

$$(a + b) + a =_{\text{AC}} b + (a + a) \overset{\text{A3}}{\to} b + a,$$
$$(b + a) + b =_{\text{AC}} a + (b + b) \overset{\text{A3}}{\to} a + b.$$

Finally, since the two normal forms are equivalent modulo AC of the +, $b + a =_{AC} a + b$, we conclude that the two original terms are provably equal.

Example 2.4.2. We equate $(a + a)(cd) + (bc)(d + d)$ and $((b + a)(c + c))d$. First, these basic process terms are reduced to normal form. In each step, the subterm that is reduced is underlined.

$$(\underline{a + a})(cd) + (bc)(d + d) \qquad\qquad ((b + a)(\underline{c + c}))d$$
$$\overset{A3}{\to} a(cd) + (bc)(\underline{d + d}) \qquad\qquad \overset{A3}{\to} ((b + a)c)d$$
$$\overset{A3}{\to} a(cd) + \underline{(bc)d} \qquad\qquad \overset{A5}{\to} \underline{(b + a)(cd)}$$
$$\overset{A5}{\to} a(cd) + b(cd) \qquad\qquad \overset{A4}{\to} b(cd) + a(cd).$$

Finally, since the two normal forms are equivalent modulo AC of the +, we conclude that the two original terms are provably equal.

Note that the reductions in the last example are not unique, because in several cases more than one subterm can be reduced. Therefore, a mechanised proof calls for a rewriting strategy, to determine which subterm is reduced by which rewrite rule. In the proof of the completeness theorem for BPA it was ensured that each of these rewriting strategies produces the same normal form from a given input term, modulo AC of the +.

Exercise 2.4.6. Derive the following three equations from $\mathcal{E}_{BPA}$:

- $((a + a)(b + b))(c + c) = a(bc)$;
- $(a + a)(bc) + (ab)(c + c) = (a(b + b))(c + c)$;
- $((a + b)c + ac)d = (b + a)(cd)$.

The axiomatisation $\mathcal{E}_{BPA}$ is ω-complete (see Definition A.3.2), meaning that if all closed instantiations of an equation can be derived from this axiomatisation, then the equation itself can be derived from this axiomatisation.

Theorem 2.4.4. *The axiomatisation $\mathcal{E}_{BPA}$ is ω-complete.*

Though Theorem 2.4.4 is independent of bisimulation equivalence, it can be proved in a similar fashion as completeness of $\mathcal{E}_{BPA}$ for BPA modulo bisimulation equivalence; see the proof of Theorem 2.4.3. The only extra is that variables need to be supplied with an operational semantics, giving rise to an extension of bisimulation equivalence to open terms (see Definition A.1.2). This extension should be such that for all open terms s and t:

(1) if $\sigma(s) \leftrightarrow \sigma(t)$ for all closed substitutions σ, then $s \leftrightarrow t$;
(2) if $s \leftrightarrow t$, then $s = t$.

Namely, by soundness of the axioms, $\sigma(s) = \sigma(t)$ implies $\sigma(s) \leftrightarrow \sigma(t)$ for all closed substitutions σ. According to (1) this yields $s \leftrightarrow t$, so by (2) $s = t$. See [2, 101, 156] for examples of this proof technique.

For the ω-completeness proof of $\mathcal{E}_{BPA}$, variables are to be interpreted as atomic actions, meaning that the transition rule $\dfrac{}{x \overset{x}{\to} \surd}$ is added to the TSS of BPA. Then (2) can be proved along the lines of the proof of Theorem 2.4.3.

3. Algebra of Communicating Processes

Atomic actions and the operators alternative and sequential composition from the previous chapter provide relatively primitive tools to construct an LTS. In general, the size of a basic process term is comparable to the size of the related process graph. This chapter introduces operators to express parallelism and concurrency, which enable us to capture a large process graph by means of a comparatively small process term.

3.1 Parallelism and Communication

In practice, process behaviour is often composed of several processors that are executed in parallel, where these separate entities may influence each other's execution. One could say that the processors are the building blocks that make up the complete system, cemented together by mutual communication actions. In order to model such concurrent systems, Milner [151] introduced the *merge*, which is a binary operator that executes the two process terms in its arguments in parallel. That is, $s\|t$ can choose to execute an initial transition of s (i.e., a transition $s \xrightarrow{a} s'$ or $s \xrightarrow{a} \checkmark$) or an initial transition of t. This is formalised by four transition rules for the merge:

$$\frac{x \xrightarrow{v} \checkmark}{x\|y \xrightarrow{v} y} \quad \frac{x \xrightarrow{v} x'}{x\|y \xrightarrow{v} x'\|y}$$

$$\frac{y \xrightarrow{v} \checkmark}{x\|y \xrightarrow{v} x} \quad \frac{y \xrightarrow{v} y'}{x\|y \xrightarrow{v} x\|y'}$$

Moreover, $s\|t$ can choose to execute a communication between initial transitions of s and t. For this purpose we assume a *communication function* $\gamma : A \times A \to A$, which produces for each pair of atomic actions a and b their communication $\gamma(a, b)$. This communication function is required to be commutative and associative; that is, for $a, b, c \in A$,

$$\gamma(a, b) \equiv \gamma(b, a)$$
$$\gamma(\gamma(a, b), c) \equiv \gamma(a, \gamma(b, c)).$$

The next four transition rules for the merge express that $s\|t$ can choose to execute a communication of initial transitions of s and t:

$$\frac{x \xrightarrow{v} \surd \quad y \xrightarrow{w} \surd}{x\|y \xrightarrow{\gamma(v,w)} \surd} \qquad \frac{x \xrightarrow{v} \surd \quad y \xrightarrow{w} y'}{x\|y \xrightarrow{\gamma(v,w)} y'}$$

$$\frac{x \xrightarrow{v} x' \quad y \xrightarrow{w} \surd}{x\|y \xrightarrow{\gamma(v,w)} x'} \qquad \frac{x \xrightarrow{v} x' \quad y \xrightarrow{w} y'}{x\|y \xrightarrow{\gamma(v,w)} x'\|y'}$$

The variables x, x', y, and y' in the eight transition rules for the merge range over the collection of process terms, while v and w range over the set A of atomic actions.

Example 3.1.1. Let the communication of two atomic actions from $\{a, b\}$ always result to c. The process graph of the process term $(ab)\|(ba)$ is depicted in Fig. 3.1.

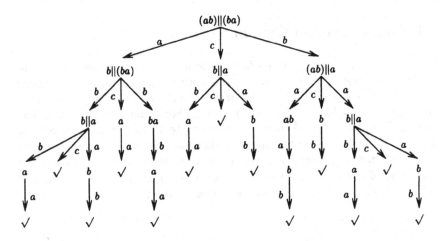

Fig. 3.1. Process graph of $(ab)\|(ba)$

Example 3.1.1 shows that the merge of two simple process terms produces a relatively large process graph. This partly explains the strength of a theory of communicating processes; in general it is much easier to study the separate components of a concurrent system than the full system itself.

Exercise 3.1.1. Let the communication of two atomic actions from $\{a, b\}$ always result to c. Find the process graph that belongs to the process term $((ab)a)\|b$. Give the derivations of the transitions in this process graph from the transition rules of BPA with the merge operator.

3.2 Left Merge and Communication Merge

Moller [159] proved that there does not exist a sound and complete finite axiomatisation for BPA extended with the merge, modulo bisimulation equivalence. This problem is overcome by defining two extra operators called *left merge* and *communication merge*, which both capture part of the behaviour of the merge. These operators were introduced by Bergstra and Klop [41], to answer an open question posed by de Bakker and Zucker [29].

The left merge $s \mathbin{\underline{\parallel}} t$ takes its initial transition from the process term s, and then behaves as the merge $\parallel$. This is expressed by two transition rules for the left merge, which correspond with the first two transition rules for the merge:

$$\frac{x \xrightarrow{v} \sqrt{}}{x \mathbin{\underline{\parallel}} y \xrightarrow{v} y} \quad \frac{x \xrightarrow{v} x'}{x \mathbin{\underline{\parallel}} y \xrightarrow{v} x' \parallel y}$$

The communication merge $s \mid t$ executes as initial transition a communication between initial transitions of the process terms s and t, and then behaves as the standard merge operator $\parallel$. This is expressed by four transition rules for the communication merge, which correspond with the last four transition rules for the merge:

$$\frac{x \xrightarrow{v} \sqrt{} \quad y \xrightarrow{w} \sqrt{}}{x \mid y \xrightarrow{\gamma(v,w)} \sqrt{}} \quad \frac{x \xrightarrow{v} \sqrt{} \quad y \xrightarrow{w} y'}{x \mid y \xrightarrow{\gamma(v,w)} y'}$$

$$\frac{x \xrightarrow{v} x' \quad y \xrightarrow{w} \sqrt{}}{x \mid y \xrightarrow{\gamma(v,w)} x'} \quad \frac{x \xrightarrow{v} x' \quad y \xrightarrow{w} y'}{x \mid y \xrightarrow{\gamma(v,w)} x' \parallel y'}$$

As binding convention we assume that the $\parallel$, $\underline{\parallel}$, and $\mid$ bind stronger than the $+$. For example, $a \mathbin{\underline{\parallel}} b + a \mid c$ represents $(a \mathbin{\underline{\parallel}} b) + (a \mid c)$. We refer to BPA extended with the three parallel operators $\parallel$, $\underline{\parallel}$, and $\mid$ as PAP (for process algebra with parallelism).

The left and communication merge together cover the behaviour of the merge, in the sense that $s \parallel t \underline{\leftrightarrow} (s \mathbin{\underline{\parallel}} t + t \mathbin{\underline{\parallel}} s) + s \mid t$ for all process terms s and t in PAP. Namely, $s \parallel t$ can execute either an initial transition of s or t, which is covered by $s \mathbin{\underline{\parallel}} t$ or $t \mathbin{\underline{\parallel}} s$, respectively, or a communication of initial transitions of s and t, which is covered by $s \mid t$. This point will be elaborated later on.

Exercise 3.2.1. Prove that the following pairs of process terms are bisimilar, for process terms s, t, and u in PAP:

- $s \parallel t$ and $(s \mathbin{\underline{\parallel}} t + t \mathbin{\underline{\parallel}} s) + s \mid t$;
- $s \parallel t$ and $t \parallel s$;
- $s \mid t$ and $t \mid s$;
- $(s \parallel t) \parallel u$ and $s \parallel (t \parallel u)$;

- $(s|t)|u$ and $s|(t|u)$;
- $(s\mathbin{\parallel} t)\mathbin{\parallel} u$ and $s\mathbin{\parallel}(t\|u)$;
- $(s\mathbin{\parallel} t)|u$ and $(s|u)\mathbin{\parallel} t$.

We want the TSS of PAP to be a conservative extension (see Definition B.5.1) of the TSS of BPA, meaning that the fourteen transition rules for the three parallel operators do not influence the process graphs of basic process terms. That is, an initial transition of a basic process term should be derivable from the TSS of PAP if and only if this transition can be derived from the TSS of BPA.

Theorem 3.2.1. *PAP is a conservative extension of BPA.*

Proof. This theorem follows from the following two facts.

1. The transition rules of BPA in Table 2.1 are all source-dependent (see Definition B.5.2).
2. The sources (see Definition B.1.2) of the fourteen transition rules for the three parallel operators all contain an occurrence of $\|$, $\mathbin{\parallel}$, or $|$.

Since the TSS of BPA is source-dependent, and the transition rules for the three parallel operators contain a fresh operator (see Definition B.5.3) in their sources, Theorem B.5.1 says that PAP is a conservative extension of BPA. $\square$

Exercise 3.2.2. Show that the transition rules of BPA are source-dependent.

PAP can only have a sound and complete axiomatisation modulo bisimulation if this equivalence is a congruence with respect to PAP. In other words, if $s \leftrightarrow s'$ and $t \leftrightarrow t'$, then it has to be the case that $s+t \leftrightarrow s'+t'$, $s\cdot t \leftrightarrow s'\cdot t'$, $s\|t \leftrightarrow s'\|t'$, $s\mathbin{\parallel} t \leftrightarrow s'\mathbin{\parallel} t'$, and $s|t \leftrightarrow s'|t'$.

Theorem 3.2.2. *Bisimulation equivalence is a congruence with respect to PAP.*

Proof. The transition rules for the three parallel operators, as well as of BPA, are all in panth format. So the bisimulation equivalence that they induce is a congruence; see Theorem B.3.1. $\square$

Exercise 3.2.3. Verify that the transition rules for the three parallel operators are in panth format.

3.3 Axioms for PAP

We are after an axiomatisation $\mathcal{E}_{\text{PAP}}$ such that the induced equality relation characterises bisimulation equivalence over PAP in the following sense:

1. $\mathcal{E}_{\text{PAP}}$ is sound, i.e., if $s = t$ can be derived from the axioms in $\mathcal{E}_{\text{PAP}}$ for certain process terms s and t in PAP, then $s \leftrightarrow t$;

2. $\mathcal{E}_{\text{PAP}}$ is complete, i.e., if $s \underleftrightarrow{} t$ holds for certain process terms s and t in PAP, then $s = t$ can be derived from the axioms in $\mathcal{E}_{\text{PAP}}$.

Table 3.1 presents the axioms for the three parallel operators modulo bisimulation equivalence. We already noted that the merge can be split into the left merge and the communication merge, in the sense that $s\|t$ is bisimilar with $(s\,\|\!\!\lfloor\, t + t\,\|\!\!\lfloor\, s) + s|t$; this is exploited in axiom M1. Axioms LM2-4 and CM5-10 enable us to eliminate occurrences of the left merge and the communication merge from process terms. The variables x, y, and z in the axioms range over process terms, while v and w range over the set A of atomic actions. The axiomatisation $\mathcal{E}_{\text{PAP}}$ consists of $\mathcal{E}_{\text{BPA}}$ together with the axioms in Table 3.1.

Table 3.1. Axioms for merge, left merge, and communication merge

M1	$x\|y = (x\,\|\!\!\lfloor\, y + y\,\|\!\!\lfloor\, x) + x|y$
LM2	$v\,\|\!\!\lfloor\, y = v \cdot y$
LM3	$(v \cdot x)\,\|\!\!\lfloor\, y = v \cdot (x\|y)$
LM4	$(x + y)\,\|\!\!\lfloor\, z = x\,\|\!\!\lfloor\, z + y\,\|\!\!\lfloor\, z$
CM5	$v|w = \gamma(v, w)$
CM6	$v|(w \cdot y) = \gamma(v, w) \cdot y$
CM7	$(v \cdot x)|w = \gamma(v, w) \cdot x$
CM8	$(v \cdot x)|(w \cdot y) = \gamma(v, w) \cdot (x\|y)$
CM9	$(x + y)|z = x|z + y|z$
CM10	$x|(y + z) = x|y + x|z$

Theorem 3.3.1. $\mathcal{E}_{\text{PAP}}$ *is sound for PAP modulo bisimulation equivalence.*

Proof. Since bisimulation is both an equivalence and a congruence, we only need to check that the first clause in the definition of the relation $=$ is sound. That is, if $s = t$ is an axiom in $\mathcal{E}_{\text{PAP}}$ and σ a closed substitution that maps the variables in s and t to process terms, then we need to check that $\sigma(s) \underleftrightarrow{} \sigma(t)$. Soundness of the axioms A1-5 can be checked as in the proof of soundness of $\mathcal{E}_{\text{BPA}}$, in Theorem 2.4.2. Here, we only provide some intuition for soundness of the axioms in Table 3.1:

- M1 is the defining axiom for the merge, which says that each initial transition of $s\|t$ stems from s (expressed by the summand $s\,\|\!\!\lfloor\, t$) or from t (expressed by the summand $t\,\|\!\!\lfloor\, s$), or is a communication of initial transitions from s and t (expressed by the summand $s|t$);
- LM2,3 are the defining axioms for the left merge, which say that $s\,\|\!\!\lfloor\, t$ takes its initial transition from s;
- LM4 (**right distributivity of** $\|\!\!\lfloor$) says that in a term $(s + t)\,\|\!\!\lfloor\, u$, a choice for an initial transition from s or t is a choice for $s\,\|\!\!\lfloor\, u$ or $t\,\|\!\!\lfloor\, u$, respectively;

- CM5-8 are the defining axioms for the communication merge, which say that $s|t$ makes as initial transition a communication of initial transitions from s and t;
- CM9 (right distributivity of $|$) says that in a term $(s+t)|u$, a choice for an initial transition from s or t is a choice for $s|u$ or $t|u$, respectively;
- CM10 (left distributivity of $|$) says that in a term $s|(t+u)$, a choice for an initial transition from t or u is a choice for $s|t$ or $s|u$, respectively.

These intuitions can be made rigorous by means of explicit bisimulation relations between the left- and right-hand sides of closed instantiations of the axioms in Table 3.1. Hence, all such instantiations are sound modulo bisimulation equivalence. $\square$

Exercise 3.3.1. Prove soundness of the axioms LM3,4 and CM8,10; that is, the following four statements are valid for actions a and b and process terms s, t, and u in PAP:

- $(as) \mathbin{\underline{\|}} t \underline{\leftrightarrow} a(s\|t)$;
- $(s+t) \mathbin{\underline{\|}} u \underline{\leftrightarrow} s \mathbin{\underline{\|}} u + t \mathbin{\underline{\|}} u$;
- $(as)|(bt) \underline{\leftrightarrow} \gamma(a,b)(s\|t)$;
- $s|(t+u) \underline{\leftrightarrow} s|t + s|u$.

Exercise 3.3.2. Give counter-examples to show that right distributivity of the merge, $(x+y)\|z = x\|z + y\|z$, and left distributivity of the left merge, $x \mathbin{\underline{\|}} (y+z) = x \mathbin{\underline{\|}} y + x \mathbin{\underline{\|}} z$, are not sound modulo bisimulation equivalence.

Exercise 3.3.3. Let t be a process term in PAP, and let

$$\{t \xrightarrow{a_i} t_i \mid i \in \{1,\ldots,k\}\} \cup \{t \xrightarrow{b_j} \sqrt{} \mid j \in \{1,\ldots,\ell\}\}$$

be the set of initial transitions of t. Prove that the equation

$$t = a_1 t_1 + \cdots + a_k t_k + b_1 + \cdots + b_\ell$$

can be derived from $\mathcal{E}_{\text{PAP}}$. (Hint: apply structural induction with respect to the size of t.)

We proceed to prove that $\mathcal{E}_{\text{PAP}}$ is complete for PAP modulo bisimulation equivalence, meaning that $s \underline{\leftrightarrow} t$ implies $s = t$. As before, the proof is based on a term rewriting analysis, in which the axioms are directed from left to right.

Theorem 3.3.2. *$\mathcal{E}_{\text{PAP}}$ is complete for PAP modulo bisimulation equivalence.*

Proof. The axioms A3-5 in $\mathcal{E}_{\text{BPA}}$ and the axioms M1, LM2-4, and CM5-10 are turned into rewrite rules, by directing them from left to right. The resulting TRS is applied to process terms in PAP modulo AC of the $+$.

The TRS is terminating modulo AC of the $+$. That is, each reduction of a process term ends up in a normal form, which cannot be reduced any further. This can be seen by defining inductively an appropriate weight function on process terms, which extends the weight function in the proof of Theorem 2.4.3 as follows:

$$weight(s\|t) \triangleq 3 \cdot (weight(s) \cdot weight(t))^2 + 1$$
$$weight(s \mathbin{\rlap{\rule[0.3em]{0.6em}{0.4pt}}{\|}} t) \triangleq (weight(s) \cdot weight(t))^2$$
$$weight(s|t) \triangleq (weight(s) \cdot weight(t))^2.$$

It is not hard to see that each application of a rewrite rule strictly decreases the weight of a process term, and that moreover process terms that are equivalent modulo AC of the $+$ have the same weight. Hence, the TRS is terminating modulo AC of the $+$.

We prove that normal forms n do not contain occurrences of the three parallel operators $\|$, $\mathbin{\rlap{\rule[0.3em]{0.6em}{0.4pt}}{\|}}$, and $|$. The proof is based on induction with respect to the size of the normal form n.

- If n is an atomic action, then it does not contain any parallel operators.
- Suppose $n =_{\text{AC}} s + t$ or $n =_{\text{AC}} s \cdot t$. Then by induction the normal forms s and t do not contain any parallel operators, so that n does not contain any parallel operators either.
- n cannot be of the form $s\|t$, because in that case the directed version of M1 would apply to it, contradicting the fact that n is a normal form.
- Suppose $n =_{\text{AC}} s \mathbin{\rlap{\rule[0.3em]{0.6em}{0.4pt}}{\|}} t$. By induction, the normal form s does not contain any parallel operators. We distinguish the possible forms of the normal form s:
  - if $s \equiv a$, then the directed version of LM2 applies to $s \mathbin{\rlap{\rule[0.3em]{0.6em}{0.4pt}}{\|}} t$;
  - if $s =_{\text{AC}} au$, then the directed version of LM3 applies to $s \mathbin{\rlap{\rule[0.3em]{0.6em}{0.4pt}}{\|}} t$;
  - if $s =_{\text{AC}} u + u'$, then the directed version of LM4 applies to $s \mathbin{\rlap{\rule[0.3em]{0.6em}{0.4pt}}{\|}} t$.
  These three cases, which cover the possible forms of the normal form s, contradict the fact that n is a normal form. We conclude that n cannot be of the form $s \mathbin{\rlap{\rule[0.3em]{0.6em}{0.4pt}}{\|}} t$.
- Suppose $n =_{\text{AC}} s|t$. By induction the normal forms s and t do not contain any parallel operators. Similar as in the previous case, we can distinguish the possible forms of s and t, which all lead to the conclusion that one of the directed versions of CM5-10 can be applied to n. We conclude that n cannot be of the form $s|t$. The analysis of the possible forms of s and t is left to the reader.

Hence, normal forms do not contain occurrences of parallel operators. In other words, normal forms are basic process terms.

We proceed to prove that the axiomatisation $\mathcal{E}_{\text{PAP}}$ is complete for PAP modulo bisimulation equivalence. Let the process terms s and t be bisimilar. The TRS is terminating modulo AC of the $+$, so it reduces s and t to normal forms n and n', respectively. Since the rewrite rules and equivalence modulo AC of the $+$ can be derived from $\mathcal{E}_{\text{PAP}}$, $s = n$ and $t = n'$. Soundness of the axioms then yields $s \underleftrightarrow{\ } n$ and $t \underleftrightarrow{\ } n'$, so $n \underleftrightarrow{\ } s \underleftrightarrow{\ } t \underleftrightarrow{\ } n'$. We showed that the normal forms n and n' are basic process terms. Then it follows, as in the proof of Theorem 2.4.3, that $n \underleftrightarrow{\ } n'$ implies $n =_{AC} n'$. Hence, $s = n =_{AC} n' = t$. $\square$

The proof of Theorem 3.3.2 points out a mechanised way to verify whether two process terms in PAP are bisimilar. First, reduce both process terms to a normal form, by means of the rewrite rules. Next, check whether the two resulting normal forms are equivalent modulo AC of the $+$. If so, then the original terms are bisimilar; if not, then the original terms are not bisimilar.

Exercise 3.3.4. Verify for the TRS in the proof of Theorem 3.3.2 that if $s \to t$ then $weight(s) > weight(t)$.

Example 3.3.1. Let the communication of two actions from $\{a, b\}$ always result to c. We show how $(ab)\|b$ is reduced to its normal form; in each step, the subterm that is reduced is underlined.

$$
\begin{aligned}
&\underline{(ab)\|b} \\
\overset{\text{M1}}{\to}\ &\underline{(ab)\, \mathbb{L}\, b} + b\, \mathbb{L}\, (ab) + (ab)|b \\
\overset{\text{LM3}}{\to}\ &a(b\|b) + \underline{b\, \mathbb{L}\, (ab)} + (ab)|b \\
\overset{\text{LM2}}{\to}\ &a(b\|b) + b(ab) + \underline{(ab)|b} \\
\overset{\text{CM7}}{\to}\ &a(\underline{b\|b}) + b(ab) + cb \\
\overset{\text{M1}}{\to}\ &a(\underline{b\, \mathbb{L}\, b + b\, \mathbb{L}\, b} + b|b) + b(ab) + cb \\
\overset{\text{A3}}{\to}\ &a(\underline{b\, \mathbb{L}\, b} + b|b) + b(ab) + cb \\
\overset{\text{LM2}}{\to}\ &a(bb + \underline{b|b}) + b(ab) + cb \\
\overset{\text{CM5}}{\to}\ &a(bb + c) + b(ab) + cb.
\end{aligned}
$$

Exercise 3.3.5. Let the communication of two actions from $\{a, b\}$ always result to c. Reduce the process term $b\|(ab)$ to its normal form. Derive the equation $(ab)\|b = b\|(ab)$ from $\mathcal{E}_{PAP}$.

Exercise 3.3.6. Derive $a\|((b + c)d) = ((b + c)d)\|a$ from $\mathcal{E}_{PAP}$.

The axiomatisation $\mathcal{E}_{PAP}$ is not ω-complete. For instance, $s\|t \underleftrightarrow{\ } t\|s$ for all process terms s and t in PAP (see the second case of Exercise 3.2.1), so according to Theorem 3.3.2, all closed substitution instances of the equation $x\|y = y\|x$ can be derived from $\mathcal{E}_{PAP}$. However, $x\|y = y\|x$ itself cannot be derived from $\mathcal{E}_{PAP}$, which follows from the fact that only the right-hand side of A3 applies to (a subterm of) $x\|y$ or $y\|x$.

Exercise 3.3.7. Derive the equations $s|t = t|s$ and $s\|t = t\|s$ from $\mathcal{E}_{PAP}$ for all process terms s and t in PAP.

Exercise 3.3.8. Give counter-examples to show that commutativity and associativity of the left merge, $x\, \mathbb{L}\, y = y\, \mathbb{L}\, x$ and $(x\, \mathbb{L}\, y)\, \mathbb{L}\, z = x\, \mathbb{L}\, (y\, \mathbb{L}\, z)$, are not sound modulo bisimulation equivalence.

3.4 Deadlock and Encapsulation

If two atomic actions are able to communicate, then often we only want these actions to occur in communication with each other, and not on their own. For example, let the action $send(d)$ represent sending a datum d into one end of a channel, while $read(d)$ represents receiving this datum at the other end of the channel. Furthermore, let the communication of these two actions result in transferring the datum d through the channel by the action $comm(d)$. For the outside world, the actions $send(d)$ and $read(d)$ never appear on their own, but only in communication in the form $comm(d)$.

In order to enforce communication in such cases, we introduce a special constant δ called *deadlock*, which does not display any behaviour. The communication function γ is extended by allowing that the communication of two atomic actions results to δ, i.e., $\gamma : A \times A \to A \cup \{\delta\}$. This extension of γ enables us to express that two actions a and b do not communicate, by defining $\gamma(a, b) \triangleq \delta$. Furthermore, we introduce unary *encapsulation* operators ∂_H for sets H of atomic actions, which rename all actions in H into δ. Deadlock and encapsulation were introduced by Milner [151]; our treatment of these notions is based on [41]. PAP extended with deadlock and encapsulation operators is called the *algebra of communicating processes* (ACP).

Since the deadlock does not display any behaviour, there is no transition rule for this constant. Furthermore, since the communication of actions can result to δ, the last four transition rules for the merge and the four transition rules for the communication merge need to be supplied with the requirement $\gamma(v, w) \not\equiv \delta$. Finally, the behaviour of the encapsulation operators is captured by the following transition rules, which express that $\partial_H(t)$ can execute all transitions of t of which the labels are not in H:

$$\frac{x \xrightarrow{v} \sqrt{}}{\partial_H(x) \xrightarrow{v} \sqrt{}} \quad v \notin H \qquad\qquad \frac{x \xrightarrow{v} x'}{\partial_H(x) \xrightarrow{v} \partial_H(x')} \quad v \notin H$$

The variables x and x' range over process terms, while v ranges over A.

Exercise 3.4.1. Verify, using the transition rules for sequential composition, left merge, and communication merge, that process terms of the form δt, $\delta \mathbin{\lfloor\lfloor} t$, $\delta | t$, and $t | \delta$ do not display any behaviour. In other words, these process terms are bisimilar to δ.

Exercise 3.4.2. Derive the process graphs of the following process terms:

- $\partial_{\{a\}}(ac)$;
- $\partial_{\{a\}}((a + b)c)$;
- $\partial_{\{c\}}((a + b)c)$;
- $\partial_{\{a,b\}}((ab)\|(ba))$ with $\gamma(a, b) = c$.

Exercise 3.4.3. Prove that the following pairs of process terms are bisimilar, for process terms s and t in ACP:

- $(s\delta)\|t$ and $(s\|t)\delta$;
- $\partial_G(\partial_H(t))$ and $\partial_{G\cup H}(t)$;
- $\partial_A(t)$ and δ;
- $\partial_\emptyset(t)$ and t, where $\emptyset$ denotes the empty set.

In Example 3.1.1 we drew the relatively large process graph of the process term $(ab)\|(ba)$, with all communications between atomic actions resulting to c. The last case in Exercise 3.4.2 shows that encapsulation can be an effective means to limit the size of the process graph of such a concurrent system. We give a further example of the use of encapsulation operators.

Example 3.4.1. Suppose a datum 0 or 1 is sent into a channel, which is expressed by the process term $send(0) + send(1)$. Let this datum be received at the other side of the channel, which is expressed by the process term $read(0) + read(1)$. The communication of $send(d)$ and $read(d)$ results to $comm(d)$ for $d \in \{0,1\}$, while all other communications between actions result to δ. The behaviour of the channel is described by the process term

$$\partial_{\{send(0),\, send(1),\, read(0),\, read(1)\}}((send(0) + send(1))\|(read(0) + read(1)))$$

The encapsulation operator enforces that the action $send(d)$ can only occur in communication with the action $read(d)$, for $d \in \{0,1\}$.

Exercise 3.4.4. Prove from the transition rules that the process term in Example 3.4.1 displays the desired behaviour of the channel; that is, it executes either $comm(0)$ or $comm(1)$, after which it terminates successfully.

Beware not to confuse a transition of the form $t \xrightarrow{a} \delta$ with a transition of the form $t \xrightarrow{a} \sqrt{}$; intuitively, the first transition expresses that t gets stuck after the execution of a, while the second transition expresses that t terminates successfully after the execution of a. A process term t is said to contain a deadlock if there are transitions $t \xrightarrow{a_1} t_1 \xrightarrow{a_2} \cdots \xrightarrow{a_3} t_n$ such that the process term t_n does not have any initial transitions (i.e., $t_n \not\rightarrow \delta$). In general it is undesirable that a process contains a deadlock, because it represents that the process gets stuck without producing any output. Experience learns that non-trivial specifications of system behaviour often contain a deadlock. For example, the third sliding window protocol in [183] contains a deadlock; see [107, *Stelling 7*]. It can, however, be very difficult to detect such a deadlock, even if one has a good insight into such a protocol. Automated tools have been developed to help with the detection of deadlocks; see Section 6.4.

Exercise 3.4.5. Let $\gamma(a,c) \triangleq \delta$ and $\gamma(b,c) \triangleq a$. Say for each of the following process terms whether it contains a deadlock:

- $\partial_{\{b\}}(ab + c)$;
- $\partial_{\{b\}}(a(b + c))$;
- $\partial_{\{b,c\}}(a(b + c))$;

- $\partial_{\{b\}}((ab)\|c)$;
- $\partial_{\{b,c\}}((ab)\|c)$.

As before, we want ACP to be a conservative extension of PAP. That is, the transition rules for the encapsulation operators should not influence the process graphs belonging to process terms in PAP.

Theorem 3.4.1. *ACP is a conservative extension of PAP.*

Proof. This theorem follows from the following two facts.

1. The transition rules of PAP are all source-dependent.
2. The sources of the transition rules for the encapsulation operators contain an occurrence of ∂_H.

Since the TSS of PAP is source-dependent, and the transition rules for encapsulation contain a fresh operator in their sources, Theorem B.5.1 says that ACP is a conservative extension of PAP. $\square$

Exercise 3.4.6. Verify that the transition rules for the parallel operators are source-dependent.

In order to be able to capture bisimulation equivalence over ACP by a sound and complete axiomatisation, it needs to be a congruence. In other words, if $s \underleftrightarrow{} s'$ and $t \underleftrightarrow{} t'$, then it has to be the case that $s + t \underleftrightarrow{} s' + t'$, $s \cdot t \underleftrightarrow{} s' \cdot t'$, $s\|t \underleftrightarrow{} s'\|t'$, $s \mathbin{\llfloor} t \underleftrightarrow{} s' \mathbin{\llfloor} t'$, $s|t \underleftrightarrow{} s'|t'$, and finally $\partial_H(s) \underleftrightarrow{} \partial_H(s')$ for all subsets H of A.

Theorem 3.4.2. *Bisimulation equivalence is a congruence with respect to ACP.*

Proof. This theorem follows from the fact that the transition rules for the encapsulation operators, as well as of PAP, are in panth format; see Theorem B.3.1. $\square$

Table 3.2 presents axioms A6,7 for the deadlock, axioms D1-5 for encapsulation, and axioms LM11 and CM12,13 to deal with the interplay of the deadlock with left and communication merge. The variables x and y range over process terms, while v ranges over A. The axioms in Table 3.2 together with $\mathcal{E}_{\mathrm{PAP}}$ are denoted by $\mathcal{E}_{\mathrm{ACP}}$.

Theorem 3.4.3. $\mathcal{E}_{\mathrm{ACP}}$ *is sound for ACP modulo bisimulation equivalence.*

Proof. Since bisimulation is both an equivalence and a congruence, we only need to check that the first clause in the definition of the relation = is sound. That is, if $s = t$ is an axiom in $\mathcal{E}_{\mathrm{ACP}}$ and σ a closed substitution that maps the variables in s and t to process terms, then we need to check that $\sigma(s) \underleftrightarrow{} \sigma(t)$. Soundness of the axioms in $\mathcal{E}_{\mathrm{PAP}}$ can be checked as before. Here, we only provide some intuition for soundness of the axioms in Table 3.2:

Table 3.2. Axioms for deadlock and encapsulation

A6		$x + \delta = x$
A7		$\delta \cdot x = \delta$
D1	$v \notin H$	$\partial_H(v) = v$
D2	$v \in H$	$\partial_H(v) = \delta$
D3		$\partial_H(\delta) = \delta$
D4		$\partial_H(x + y) = \partial_H(x) + \partial_H(y)$
D5		$\partial_H(x \cdot y) = \partial_H(x) \cdot \partial_H(y)$
LM11		$\delta \mathbin{\Vert\mkern-6mu\raise1pt\hbox{$_$}} x = \delta$
CM12		$\delta \mid x = \delta$
CM13		$x \mid \delta = \delta$

- A6 says that the deadlock δ displays no behaviour, so that in a process term $t + \delta$ the summand δ is redundant;
- A7, LM11, and CM12,13 say that the deadlock δ blocks all behaviour, so that process terms δt, $\delta \mathbin{\Vert\mkern-6mu\raise1pt\hbox{$_$}} t$, $\delta \mid t$, and $t \mid \delta$ do not display any behaviour (see Exercise 3.4.1);
- D1-3 are the defining equations for the encapsulation operator ∂_H: D2 says that it renames atomic actions from H into δ, while D1,3 say that it leaves atomic actions outside H and the deadlock δ unchanged;
- D4,5 say that in $\partial_H(t)$, all transitions of t labelled with atomic actions from H are blocked.

These intuitions can be made rigorous by means of explicit bisimulation relations between the left- and right-hand sides of closed instantiations of the axioms in Table 3.2. Hence, all such instantiations are sound modulo bisimulation equivalence. □

Exercise 3.4.7. Give a counter-example to show that the equation $\partial_H(x \| y) = \partial_H(x) \| \partial_H(y)$ is not sound modulo bisimulation equivalence.

Theorem 3.4.4. $\mathcal{E}_{\mathrm{ACP}}$ *is complete for ACP modulo bisimulation equivalence.*

Proof. The axioms A6,7, D1-5, LM11, and CM12,13 are turned into rewrite rules, directed from left to right, and added to the thirteen rewrite rules for PAP in the proof of Theorem 3.3.2. The resulting TRS is terminating modulo AC of the $+$, which can be seen by inductively defining an appropriate weight function on process terms, which extends the weight function in the proof of Theorem 3.3.2 as follows:

$$weight(\delta) \triangleq 2$$
$$weight(\partial_H(s)) \triangleq 2^{weight(s)}.$$

It is not hard to see that each application of a rewrite rule strictly decreases the weight of a process term, and that moreover process terms that are equivalent modulo AC of the + have the same weight. Hence, the TRS is terminating modulo AC of the +.

As in the proof of Theorem 3.3.2, it can be shown that normal forms do not contain occurrences of the three parallel operators $\|$, $\mathbb{L}$, and $|$. We proceed to show that normal forms are not of the form $\partial_H(s)$. This fact is proved by an analysis of the possible forms of s, where we may assume that s is a normal form that does not contain occurrences of encapsulation operators:

- if $s \equiv a$, then the directed version of D1 or D2 applies to $\partial_H(s)$;
- if $s \equiv \delta$, then the directed version of D3 applies to $\partial_H(s)$;
- if $s =_{\text{AC}} t + t'$, then the directed version of D4 applies to $\partial_H(s)$;
- if $s =_{\text{AC}} tt'$, then the directed version of D5 applies to $\partial_H(s)$.

These four cases, which cover the possible forms of the normal form s, all lead to the conclusion that $\partial_H(s)$ is not a normal form. Hence, normal forms are process terms in BPA extended with the deadlock.

We proceed to prove that the axiomatisation $\mathcal{E}_{\text{ACP}}$ is complete for ACP modulo bisimulation equivalence. Let the process terms s and t be bisimilar. The TRS is terminating modulo AC of the +, so it reduces s and t to normal forms n and n', respectively. Since the rewrite rules and equivalence modulo AC of the + can be derived from $\mathcal{E}_{\text{ACP}}$, $s = n$ and $t = n'$. Soundness of the axioms then yields $s \leftrightarrow n$ and $t \leftrightarrow n'$, so $n \leftrightarrow s \leftrightarrow t \leftrightarrow n'$. We showed that the normal forms n and n' are basic process terms with possible occurrences of deadlocks. Then it follows, as in the proof of Theorem 2.4.3, that $n \leftrightarrow n'$ implies $n =_{\text{AC}} n'$. Hence, $s = n =_{\text{AC}} n' = t$. $\square$

The proof of Theorem 3.4.4 points out a mechanised way to verify whether two process terms in ACP are bisimilar. First, reduce both process terms to a normal form, by means of the rewrite rules. Next, check whether the two resulting normal forms are equivalent modulo AC of the +. If so, then the original terms are bisimilar; if not, then the original terms are not bisimilar.

Exercise 3.4.8. Prove for the TRS in the proof of Theorem 3.4.4 that if $s \to t$ then $weight(s) > weight(t)$.

Exercise 3.4.9. Suppose $s + t = \delta$ can be derived from $\mathcal{E}_{\text{ACP}}$ for certain process terms s and t in ACP. Derive $s = \delta$ from $\mathcal{E}_{\text{ACP}}$.

Exercise 3.4.10. Reduce the following process terms to their respective normal forms;

- $\delta \| a$;
- $\partial_{\{a,b\}}((ab)\|(ba))$ with $\gamma(a, b) = c$ (cf. the fourth case of Exercise 3.4.2);
- $\partial_{\{send(0),\, send(1),\, read(0),\, read(1)\}}((send(0)+send(1))\|(read(0)+read(1)))$ (cf. Example 3.4.1).

Exercise 3.4.11. Let the binary operator *alt* alternately execute an atomic action from its first and second argument. That is, the transition rules for *alt* are:

$$\frac{x \xrightarrow{v} \sqrt{}}{alt(x,y) \xrightarrow{v} y} \qquad \frac{x \xrightarrow{v} x'}{alt(x,y) \xrightarrow{v} alt(y,x')}$$

Add this operator to ACP, give axioms for the operator *alt*, and argue why they are sound modulo bisimulation equivalence.

Explain why it is possible to eliminate all occurrences of *alt* from process terms in ACP extended with *alt*, using your axioms together with $\mathcal{E}_{\mathrm{ACP}}$. Finally, show that this axiomatisation is complete for ACP with the *alt* operator modulo bisimulation equivalence.

4. Recursion

Up to now we have focussed on finite processes. However, systems can often exhibit unlimited behaviour. In this chapter it is shown how such infinite behaviour can be specified using recursive equations. For an exposition on alternative, iterative operators to express infinite behaviour, see [37].

4.1 Guarded Recursive Specifications

Consider the process that alternately executes actions a and b until infinity, with the root node presented at the top:

Since ACP can only specify finite behaviour, there does not exist a process term in ACP with this (or a bisimilar) process graph. Intuitively, the process above can be captured by means of two recursive equations:

$$X = aY$$
$$Y = bX.$$

Here, X and Y are *recursion variables*, which intuitively represent the two states of the process in which it is going to execute a or b, respectively.

Definition 4.1.1 (Recursive specification). *A recursive specification is a finite set of recursive equations*

$$X_1 = t_1(X_1, \ldots, X_n)$$
$$\vdots$$
$$X_n = t_n(X_1, \ldots, X_n)$$

where the left-hand sides X_i are recursion variables, and the right-hand sides $t_i(X_1, \ldots, X_n)$ are process terms in ACP with possible occurrences of the recursion variables $X_1, \ldots, X_n$.

Definition 4.1.2 (Solution). *Processes* $p_1, \ldots, p_n$ *are a solution for a recursive specification* $\{X_i = t_i(X_1, \ldots, X_n) \mid i \in \{1, \ldots, n\}\}$ (with respect to bisimulation equivalence) *if* $p_i \underline{\leftrightarrow} t_i(p_1, \ldots, p_n)$ *for* $i \in \{1, \ldots, n\}$.

A recursive specification should represent a unique process, so we want its solution to be unique, modulo bisimulation equivalence. That is, if $p_1, \ldots, p_n$ and $q_1, \ldots, q_n$ are two solutions for the same recursive specification, then $p_i \underline{\leftrightarrow} q_i$ for $i \in \{1, \ldots, n\}$. However, there exist recursive specifications that allow more than one solution modulo bisimulation equivalence. We give some examples.

Example 4.1.1. Let $a \in A$.

1. All processes are a solution for the recursive specification $\{X=X\}$.
2. All processes p that can execute an initial transition $p \xrightarrow{a} \surd$ are a solution for the recursive specification $\{X=a+X\}$.
3. All processes that cannot terminate successfully are a solution for the recursive specification $\{X=Xa\}$.

Exercise 4.1.1. Give two solutions for the recursive specification $\{X=a\|X\}$ that are not bisimilar.

The following example features recursive specifications that do have a unique solution modulo bisimulation equivalence.

Example 4.1.2. Let $a, b \in A$.

1. The only solution for $\{X=aY, Y=bX\}$, modulo bisimulation equivalence, is $X \underline{\leftrightarrow} abab\cdots$ and $Y \underline{\leftrightarrow} baba\cdots$.
2. The only solution for $\{X=Y, Y=aX\}$, modulo bisimulation equivalence, is $X \underline{\leftrightarrow} aaa\cdots$ and $Y \underline{\leftrightarrow} aaa\cdots$.
3. The only solution for $\{X=(a+b)\underline{\|}\, X\}$, modulo bisimulation equivalence, is $X \underline{\leftrightarrow} (a+b)(a+b)(a+b)\cdots$.

A recursive specification allows a unique solution modulo bisimulation equivalence if and only if it is *guarded*.

Definition 4.1.3 (Guarded recursive specification). *A recursive specification*

$$X_1 = t_1(X_1, \ldots, X_n)$$
$$\vdots$$
$$X_n = t_n(X_1, \ldots, X_n)$$

is guarded if the right-hand sides of its recursive equations can be adapted to the form

$$a_1 \cdot s_1(X_1, \ldots, X_n) + \cdots + a_k \cdot s_k(X_1, \ldots, X_n) + b_1 + \cdots + b_\ell$$

with $a_1, \ldots, a_k, b_1, \ldots, b_\ell \in A$, *by applications of the axioms in $\mathcal{E}_{ACP}$ and replacing recursion variables by the right-hand sides of their recursive equations. The sum above is allowed to be empty (i.e., k and ℓ can both be zero), in which case it represents the deadlock δ.*

The recursive specifications in Example 4.1.1 are all unguarded; that is, their right-hand sides cannot be brought into the desired form presented in Definition 4.1.3.

Exercise 4.1.2. Show that the recursive specifications in Example 4.1.2 are guarded.

Exercise 4.1.3. Show that $\{X=Y\|Z, Y=Z+a, Z=bZ\}$, with all communications between actions from $\{a, b\}$ resulting to c, is guarded.

4.2 Transition Rules for Guarded Recursion

If E is a guarded recursive specification, and X a recursion variable in E, then intuitively $\langle X|E\rangle$ denotes the process that has to be substituted for X in the solution for E. For instance, if E is $\{X=aY, Y=bX\}$, then $\langle X|E\rangle$ represents the process $abab\cdots$, while $\langle Y|E\rangle$ represents the process $baba\cdots$; see the first recursive specification in Example 4.1.2. We extend ACP with the constants $\langle X|E\rangle$ for guarded recursive specifications E and recursion variables X in E.

Assume that the guarded recursive specification E is of the form

$$X_1 = t_1(X_1, \ldots, X_n)$$
$$\vdots$$
$$X_n = t_n(X_1, \ldots, X_n).$$

The TSS of ACP with guarded recursion is obtained by extending the TSS of ACP with two transition rules from [98], which express that the behaviour of the solutions $\langle X_i|E\rangle$ for the recursion variables X_i in E, for $i \in \{1, \ldots, n\}$, is exactly the behaviour of their right-hand sides $t_i(X_1, \ldots, X_n)$:

$$\frac{t_i(\langle X_1|E\rangle, \ldots, \langle X_n|E\rangle) \xrightarrow{v} \sqrt{}}{\langle X_i|E\rangle \xrightarrow{v} \sqrt{}} \qquad \frac{t_i(\langle X_1|E\rangle, \ldots, \langle X_n|E\rangle) \xrightarrow{v} y}{\langle X_i|E\rangle \xrightarrow{v} y}$$

The variable y ranges over process terms, while v ranges over A.

Example 4.2.1. Let $E \triangleq \{X=aY, Y=bX\}$. The process graph of $\langle X|E\rangle$ is

$$\langle X|E\rangle$$
$$b \;\;\binom{}{}\; a$$
$$\langle Y|E\rangle$$

The transition $\langle X|E\rangle \xrightarrow{a} \langle Y|E\rangle$ can be derived from the TSS of ACP with guarded recursion as follows:

$$\frac{a \xrightarrow{a} \checkmark}{a\langle Y|E\rangle \xrightarrow{a} \langle Y|E\rangle} \quad (\frac{v \xrightarrow{v} \checkmark}{v \xrightarrow{v} \checkmark}, \quad v := a)$$

$$\frac{}{a\langle Y|E\rangle \xrightarrow{a} \langle Y|E\rangle} \quad (\frac{x \xrightarrow{v} \checkmark}{xy \xrightarrow{v} y}, \quad v := a, \ x := a, \ y := \langle Y|E\rangle)$$

$$\langle X|E\rangle \xrightarrow{a} \langle Y|E\rangle \quad (\frac{a\langle Y|E\rangle \xrightarrow{v} y}{\langle X|E\rangle \xrightarrow{v} y}, v := a, \ y := \langle Y|E\rangle)$$

Exercise 4.2.1. Derive the transition $\langle Y|E\rangle \xrightarrow{b} \langle X|E\rangle$ from the transition rules, for the guarded recursive specification E in Example 4.2.1.

From now on, for notational convenience, terms are often considered modulo associativity of sequential composition (i.e., modulo axiom A5).

Exercise 4.2.2. Derive the process graphs that belong to the following four process terms from the transition rules:

- $\langle X \mid X{=}ab\rangle$;
- $\langle X \mid X{=}YX, Y{=}bY\rangle$;
- $\langle X \mid X{=}aXb\rangle$;
- $\langle X \mid X{=}aXb{+}c\rangle$.

Theorem 4.2.1. *ACP with guarded recursion is a conservative extension of ACP.*

Proof. This theorem follows from the following two facts.

1. The transition rules of ACP are all source-dependent.
2. The sources of the transition rules for guarded recursion are of the form $\langle X|E\rangle$.

Since the TSS of ACP is source-dependent, and the sources of the transition rules for guarded recursion consist of a fresh constant, Theorem B.5.1 says that ACP with guarded recursion is a conservative extension of ACP. $\square$

Theorem 4.2.2. *Bisimulation equivalence is a congruence with respect to ACP with guarded recursion.*

Proof. This theorem follows from the fact that the transition rules for guarded recursion, as well as of ACP, are all in panth format; see Theorem B.3.1. $\square$

As an example of the use of guarded recursion we consider the bag process over the set $\{0, 1\}$; this example stems from [42] (see also [28]).

Example 4.2.2. We specify a process that can put elements 0 and 1 into a bag, and subsequently collect these elements from the bag in arbitrary order. The actions *in*(0) and *in*(1) represent putting a 0 or 1 into the bag, respectively. Similarly, the actions *out*(0) and *out*(1) represent collecting a 0 or 1 from the bag, respectively. All communications between actions result to δ. Initially the bag is empty, so that one can only put an element into the bag. The process graph in Fig. 4.1 depicts the behaviour of the bag over $\{0,1\}$, with the root state placed in the leftmost uppermost corner. Note that this bag process consists of infinitely many non-bisimilar states.

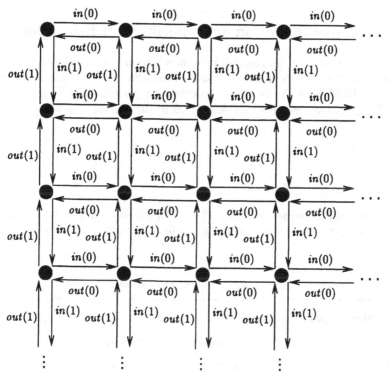

Fig. 4.1. Process graph of the bag over $\{0,1\}$

The bag over $\{0,1\}$ can be specified by a single recursive equation, using the merge $\|$. Namely, let E denote the guarded recursive specification

$$X = in(0) \cdot (X \| out(0)) + in(1) \cdot (X \| out(1)).$$

The process graph of $\langle X|E \rangle$ is bisimilar with the behaviour of the bag over $\{0,1\}$ as depicted above. Namely, initially $\langle X|E \rangle$ can only execute an action $in(d)$ for $d \in \{0,1\}$. The subsequent process term $\langle X|E \rangle \| out(d)$ can put elements 0 and 1 in the bag and take them out again (by means of the parallel component $\langle X|E\rangle$), or it can at any time take the initial element d out of the bag (by means of the parallel component $out(d)$).

Exercise 4.2.3. Give a bisimulation relation that relates the root state of the bag over $\{0,1\}$ in Fig. 4.1 and the process term $\langle X|E\rangle$ in Example 4.2.2.

Exercise 4.2.4. Suppose it would be allowed to use infinitely many recursion variables. Give a guarded recursive specification of the bag over $\{0,1\}$ that consists of infinitely many recursive equations, without using the three parallel operators.

4.3 Recursive Definition and Specification Principles

As before, we want to fit guarded recursion into an axiomatic framework. Table 4.1 contains two axioms for guarded recursion, the *recursive definition principle* (RDP) and the *recursive specification principle* (RSP) from [44, 154]. The guarded recursive specification E in the axioms is assumed to be of the form

$$X_1 = t_1(X_1, \ldots, X_n)$$
$$\vdots$$
$$X_n = t_n(X_1, \ldots, X_n).$$

Intuitively, RDP expresses that $\langle X_1|E\rangle, \ldots, \langle X_n|E\rangle$ is a solution for E, while RSP expresses that this is the only solution for E modulo bisimulation equivalence.

Table 4.1. Recursive definition and specification principles

RDP	$\langle X_i|E\rangle = t_i(\langle X_1|E\rangle, \ldots, \langle X_n|E\rangle) \qquad (i \in \{1, \ldots, n\})$
RSP	If $y_i = t_i(y_1, \ldots, y_n)$ for $i \in \{1, \ldots, n\}$, then
	$\qquad y_i = \langle X_i|E\rangle \qquad (i \in \{1, \ldots, n\})$

Theorem 4.3.1. $\mathcal{E}_{\mathrm{ACP}}+\mathrm{RDP}, \mathrm{RSP}$ *is sound for ACP with guarded recursion modulo bisimulation equivalence.*

Proof. Since bisimulation is both an equivalence and a congruence, we only need to check that if $s = t$ is an axiom in $\mathcal{E}_{\mathrm{ACP}} + \mathrm{RDP}, \mathrm{RSP}$ and σ a closed substitution that maps the variables in s and t to process terms, then $\sigma(s) \underline{\leftrightarrow} \sigma(t)$. Soundness of the axioms in $\mathcal{E}_{\mathrm{ACP}}$ can be checked as before. Here, we only provide some intuition for soundness of RDP and RSP:

- soundness of RDP follows immediately from the two transition rules for guarded recursion, which express that $\langle X_i|E\rangle$ and $t_i(\langle X_1|E\rangle,\dots,\langle X_n|E\rangle)$ have the same initial transitions for $i \in \{1,\dots,n\}$.
- soundness of RSP follows from the fact that guarded recursive specifications have only one solution modulo bisimulation equivalence.

These intuitions can be made rigorous by means of explicit bisimulation relations between the left- and right-hand sides of RDP and closed instantiations of RSP. □

Exercise 4.3.1. Give a counter-example to show that RSP is not sound for unguarded recursive specifications.

Exercise 4.3.2. Give a counter-example to show that the equation $\langle X|E\rangle \, \rotatebox[origin=c]{180}{$\sqsubset$}\!\!\!\!\rotatebox{0}{$\sqsubset$}\, y = \langle X|E\rangle \, y$, for guarded recursive specifications E, is not sound modulo bisimulation equivalence.

Example 4.3.1. The bisimilar process terms $\langle X \mid X{=}aY, Y{=}aZ, Z{=}aX\rangle$ and $\langle W \mid W{=}aaW\rangle$ can be equated by means of the axioms. This derivation consists of equating both process terms to the process term $\langle V \mid V{=}aV\rangle$.

$$\langle V \mid V{=}aV\rangle \stackrel{\text{RDP}}{=} a\langle V \mid V{=}aV\rangle.$$

Hence, substituting $\langle V \mid V{=}aV\rangle$ for the recursion variables X, Y, and Z in $\{X{=}aY, Y{=}aZ, Z{=}aX\}$ is a solution for this guarded recursive specification. So by RSP,

$$\langle V \mid V{=}aV\rangle \;=\; \langle X \mid X{=}aY, Y{=}aZ, Z{=}aX\rangle.$$

Furthermore,

$$\langle V \mid V{=}aV\rangle \stackrel{\text{RDP}}{=} a\langle V \mid V{=}aV\rangle \stackrel{\text{RDP}}{=} aa\langle V \mid V{=}aV\rangle.$$

Hence, substituting $\langle V \mid V{=}aV\rangle$ for the recursion variable W in $\{W{=}aaW\}$ is a solution for this guarded recursive specification. So by RSP,

$$\langle V \mid V{=}aV\rangle \;=\; \langle W \mid W{=}aaW\rangle.$$

Hence,

$$\langle X \mid X{=}aY, Y{=}aZ, Z{=}aX\rangle \;=\; \langle V \mid V{=}aV\rangle \;=\; \langle W \mid W{=}aaW\rangle.$$

Exercise 4.3.3. Derive the following equations from $\mathcal{E}_{\text{ACP}} + \text{RDP}, \text{RSP}$:

- $\langle X \mid X{=}aX{+}b\rangle = \langle Y \mid Y{=}aY{+}b\rangle$;
- $\langle X \mid X{=}aX\rangle = \langle Y_1 \mid Y_1{=}aY_2, Y_2{=}aY_1\rangle$;
- $\langle X \mid X{=}aaX\rangle = \langle Y \mid Y{=}aaaY\rangle$;
- $\langle X \mid X{=}aX{+}b(a{+}b)X\rangle = \langle Y \mid Y{=}bY{+}a(a{+}b)Y\rangle$;

- $\langle X \mid X{=}aX \rangle \| \langle Y \mid Y{=}bY \rangle = \langle Z \mid Z{=}(a{+}b{+}\gamma(a,b))Z \rangle$;
- $\langle X \mid X{=}aX{+}b \rangle \cdot \langle Y \mid Y{=}(a{+}b)Y \rangle = \langle Z \mid Z{=}(a{+}b)Z \rangle$;
- $\langle X \mid X{=}aX \rangle = \langle X \mid X{=}aX \rangle b$;
- $\langle X \mid X{=}aX \rangle = \langle Y \mid Y{=}aYb \rangle$.

Exercise 4.3.4. Let t_1, t_2, and t_3 be process terms with $t_1 = a(t_2b + c)$, $t_2 = ct_2 + bt_3$, and $t_3 = a(t_1 + t_3)t_2$. Prove t_1, t_2, and t_3 equal to process terms of the form $\langle X|E \rangle$, for some guarded recursive specification E.

Exercise 4.3.5. Let t_1 and t_2 be two process terms with $t_1 = at_2$ and $t_2 = at_1$. Derive the equation $t_1 = \langle X \mid X{=}aX \rangle$.

In the current framework it is not allowed to apply axioms to right-hand sides of recursive equations directly, but only indirectly using RDP and RSP. For example, consider the bisimilar process terms $\langle X \mid X{=}aX{+}aX \rangle$ and $\langle X \mid X{=}aX \rangle$. They can be equated using A3 in conjunction with RDP and RSP:

$$\langle X \mid X{=}aX{+}aX \rangle \overset{\text{RDP}}{=} a \cdot \langle X \mid X{=}aX{+}aX \rangle + a \cdot \langle X \mid X{=}aX{+}aX \rangle$$
$$\overset{\text{A3}}{=} a \cdot \langle X \mid X{=}aX{+}aX \rangle,$$

so

$$\langle X \mid X{=}aX{+}aX \rangle \overset{\text{RSP}}{=} \langle X \mid X{=}aX \rangle.$$

It is tempting, however, to conclude that these process terms are equal by a direct application of A3 with respect to the right-hand side of the recursive equation $X = aX + aX$ in the first process term. Although in principle such an application would be illegal, the following result purports the soundness of applications of axioms to right-hand sides of recursive equations. At the same time, Theorem 4.3.2 justifies the manipulation of right-hand sides of recursive equations in the definition of guarded recursive specifications; see Definition 4.1.3.

Theorem 4.3.2. *Let E_1 and E_2 be guarded recursive specifications, where E_2 is obtained from E_1 by adapting the right-hand sides of its recursive equations, using the axioms in $\mathcal{E}_{\text{ACP}}$ and the possibility to replace recursion variables by the right-hand sides of their recursive equations. Then $\langle X|E_1 \rangle = \langle X|E_2 \rangle$ can be derived from $\mathcal{E}_{\text{ACP}} + \text{RDP}, \text{RSP}$ for all recursion variables X in E_1.*

Proof. Let E_1 consist of recursive equations $X_i = t_i(X_1, \ldots, X_n)$ for $i \in \{1, \ldots, n\}$. Fix a $j \in \{1, \ldots, n\}$; we distinguish the two possible adaptations of the right-hand side of $X_j = t_j(X_1, \ldots, X_n)$ in E_1.

1. Suppose $t_j(X_1, \ldots, X_n) = s_j(X_1, \ldots, X_n)$ can be derived from $\mathcal{E}_{\text{ACP}}$. Let E_2 be obtained from E_1 by adapting the right-hand side of the j-th recursive equation in E_1 to $s_j(X_1, \ldots, X_n)$.

$$\langle X_i | E_1 \rangle \stackrel{\mathrm{RDP}}{=} t_i(\langle X_1 | E_1 \rangle, \dots, \langle X_n | E_1 \rangle) \text{ for } i \neq j$$
$$\langle X_j | E_1 \rangle \stackrel{\mathrm{RDP}}{=} t_j(\langle X_1 | E_1 \rangle, \dots, \langle X_n | E_1 \rangle)$$
$$= s_j(\langle X_1 | E_1 \rangle, \dots, \langle X_n | E_1 \rangle).$$

So replacing X_i by $\langle X_i | E_1 \rangle$ for $i \in \{1, \dots, n\}$ is a solution for E_2. Hence, by RSP, $\langle X_i | E_1 \rangle = \langle X_i | E_2 \rangle$ for $i \in \{1, \dots, n\}$.

2. Suppose $s_j(X_1, \dots, X_n)$ is the result of replacing an occurrence of X_k in $t_j(X_1, \dots, X_n)$ by $t_k(X_1, \dots, X_n)$, for some $k \in \{1, \dots, n\}$. Let E_2 be obtained from E_1 by adapting the right-hand side of the j-th recursive equation in E_1 to $s_j(X_1, \dots, X_n)$.

$$\langle X_i | E_1 \rangle \stackrel{\mathrm{RDP}}{=} t_i(\langle X_1 | E_1 \rangle, \dots, \langle X_n | E_1 \rangle) \text{ for } i \neq j$$
$$\langle X_j | E_1 \rangle \stackrel{\mathrm{RDP}}{=} t_j(\langle X_1 | E_1 \rangle, \dots, \langle X_n | E_1 \rangle)$$
$$\stackrel{\mathrm{RDP}}{=} s_j(\langle X_1 | E_1 \rangle, \dots, \langle X_n | E_1 \rangle).$$

So replacing X_i by $\langle X_i | E_1 \rangle$ for $i \in \{1, \dots, n\}$ is a solution for E_2. Hence, by RSP, $\langle X_i | E_1 \rangle = \langle X_i | E_2 \rangle$ for $i \in \{1, \dots, n\}$.

Since $=$ is closed under transitivity, the two cases above together yield the desired result. $\square$

Exercise 4.3.6. Let t be a process term in ACP with guarded recursion, and let

$$\{t \stackrel{a_i}{\rightarrow} t_i \mid i \in \{1, \dots, k\}\} \cup \{t \stackrel{b_j}{\rightarrow} \sqrt{} \mid j \in \{1, \dots, \ell\}\}$$

be the set of initial transitions of t. Prove that the equation

$$t = a_1 t_1 + \cdots + a_k t_k + b_1 + \cdots + b_\ell$$

can be derived from $\mathcal{E}_{\mathrm{ACP}} + \mathrm{RDP}$.

4.4 Completeness for Regular Processes

ACP with guarded recursion does not allow a straightforward complete axiomatisation modulo bisimulation equivalence. In particular, the axiomatisation $\mathcal{E}_{\mathrm{ACP}} + \mathrm{RDP}, \mathrm{RSP}$ is incomplete for ACP with guarded recursion. For instance, the following two symmetric guarded recursive specifications of the bag over $\{0, 1\}$ (see Example 4.2.2) are bisimilar, but cannot be proved equal by means of $\mathcal{E}_{\mathrm{ACP}} + \mathrm{RDP}, \mathrm{RSP}$:

$$X = in(0) \cdot (X \| out(0)) + in(1) \cdot (X \| out(1))$$

$$Y = in(0) \cdot (out(0) \| Y) + in(1) \cdot (out(1) \| Y).$$

(In this particular case, this could be remedied by adding a commutativity axiom for the merge.)

In this section it is shown that $\mathcal{E}_{\mathrm{ACP}} + \mathrm{RDP}, \mathrm{RSP}$ is complete for the subclass of *linear* recursive specifications.

Definition 4.4.1 (Linear recursive specification). *A recursive specification is linear if its recursive equations are of the form*

$$X \; = \; a_1 X_1 + \cdots + a_k X_k + b_1 + \cdots + b_\ell$$

with $a_1, \ldots, a_k, b_1, \ldots, b_\ell \in A$. (The empty sum represents δ.)

A regular process, which can reach only finitely many states from its root state (see Definition B.3.1), can always be described by a linear recursive specification. Namely, each reachable state s in the regular process can be represented by a recursion variable X_s. If state s can evolve into state s' by the execution of an action a, then this is expressed by a summand $aX_{s'}$ at the right-hand side of the recursive equation for X_s. Moreover, if state s can terminate successfully by the execution of an action a, then this is expressed by a summand a at the right-hand side of the recursive equation for X_s. The result is a linear recursive specification E, and $\langle X_s | E \rangle \; \underline{\leftrightarrow} \; s$ for all states s in the regular process. Vice versa, a linear recursive specification always gives rise to a regular process. Note that a linear recursive specification is by default guarded.

Exercise 4.4.1. Give a linear recursive specification E such that the regular process graph $\{s_0 \stackrel{a}{\rightarrow} s_0, s_0 \stackrel{b}{\rightarrow} s_1, s_1 \stackrel{c}{\rightarrow} s_0, s_1 \stackrel{a}{\rightarrow} s_1\}$, with root state s_0, is bisimilar to $\langle X | E \rangle$ for some recursion variable X in E.

Exercise 4.4.2. Prove that each process term in ACP with linear recursion produces a regular process graph. (Hint: apply structural induction with respect to term size.)

We prove completeness of the axiomatisation $\mathcal{E}_{\mathrm{ACP}} + \mathrm{RDP}, \mathrm{RSP}$ for ACP with linear recursive specifications. This completeness result originates from [42, 154].

Theorem 4.4.1. *$\mathcal{E}_{\mathrm{ACP}} + \mathrm{RDP}, \mathrm{RSP}$ is complete for ACP with linear recursion modulo bisimulation equivalence.*

Proof. As a first step we note that each process term t_1 in ACP with linear recursion is provably equal to a process term $\langle X_1 | E \rangle$ with E a linear recursive specification. Namely, each such process term t_1 generates a regular process graph (see Exercise 4.4.2), with states say $t_1, \ldots, t_n$. This process graph can be expressed in the form of equations

$$t_i = a_{i1} t_{i_1} + \cdots + a_{ik_i} t_{ik_i} + b_{i1} + \cdots + b_{il_i}$$

for $i \in \{1, \ldots, n\}$ (see Exercise 4.3.6). Let the linear recursive specification E consist of the recursive equations

$$X_i = a_{i1} X_{i_1} + \cdots + a_{ik_i} X_{ik_i} + b_{i1} + \cdots + b_{il_i}$$

for $i \in \{1, \ldots, n\}$. Since replacing X_i by t_i for $i \in \{1, \ldots, n\}$ is a solution for E, RSP yields $t_1 = \langle X_1 | E \rangle$.

It remains to prove that if $\langle X_1 | E_1 \rangle \leftrightarrow \langle Y_1 | E_2 \rangle$ for linear recursive specifications E_1 and E_2, then $\langle X_1 | E_1 \rangle = \langle Y_1 | E_2 \rangle$. Let E_1 and E_2 consist of recursive equations $X = t_X$ for $X \in \mathcal{X}$ and $Y = t_Y$ for $Y \in \mathcal{Y}$, respectively. The linear recursive specification E is defined to consist of the recursive equations $Z_{XY} = t_{XY}$ for $X \in \mathcal{X}$ and $Y \in \mathcal{Y}$ with $\langle X | E_1 \rangle \leftrightarrow \langle Y | E_2 \rangle$, where t_{XY} consists of the following summands:

1. t_{XY} contains a summand $a Z_{X'Y'}$ if and only if t_X and t_Y contain the summands aX' and aY', respectively, and $\langle X' | E_1 \rangle \leftrightarrow \langle Y' | E_2 \rangle$;
2. t_{XY} contains a summand b if and only if both t_X and t_Y contain the summand b.

Let the substitutions σ and ψ from recursion variables to process terms be defined as follows:

- σ maps recursion variables X in E_1 to $\langle X | E_1 \rangle$;
- ψ maps recursion variables Z_{XY} in E to $\langle X | E_1 \rangle$.

We proceed to show that substituting $\langle X | E_1 \rangle$ for recursion variables Z_{XY} in E is a solution for E; that is, $\langle X | E_1 \rangle = \psi(t_{XY})$ for recursion variables Z_{XY} in E.

Consider a recursion variable Z_{XY} in E. Then $\langle X | E_1 \rangle \leftrightarrow \langle Y | E_2 \rangle$, so for each summand aX' of t_X there is a summand aY' of t_Y with $\langle X' | E_1 \rangle \leftrightarrow \langle Y' | E_2 \rangle$. Moreover, each summand b of t_X is also a summand of t_Y. Then the definition of t_{XY} yields that for each summand aX' or b of t_X there is a summand $a Z_{X'Y'}$ or b of t_{XY}. Vice versa, if $a Z_{X'Y'}$ or b is a summand of t_{XY}, then according to the definition of t_{XY}, aX' or b is a summand of t_X. Since $\sigma(aX') \equiv a\langle X' | E_1 \rangle \equiv \psi(a Z_{X'Y'})$ and $\sigma(b) \equiv b \equiv \psi(b)$, it follows that $\sigma(t_X)$ and $\psi(t_{XY})$ consist of the same summands. So we can apply A3 to derive $\sigma(t_X) = \psi(t_{XY})$. Hence,

$$\langle X | E_1 \rangle \stackrel{\mathrm{RDP}}{=} \sigma(t_X) = \psi(t_{XY}).$$

We conclude from the derivation above that substituting process terms $\langle X | E_1 \rangle$ for recursion variables Z_{XY} in E is a solution for E. Then RSP yields $\langle X | E_1 \rangle = \langle Z_{XY} | E \rangle$ for recursion variables Z_{XY} in E, so in particular $\langle X_1 | E_1 \rangle = \langle Z_{X_1 Y_1} | E \rangle$. Likewise we can derive $\langle Y_1 | E_2 \rangle = \langle Z_{X_1 Y_1} | E \rangle$. Hence,

$$\langle X_1 | E_1 \rangle = \langle Z_{X_1 Y_1} | E \rangle = \langle Y_1 | E_2 \rangle.$$

Finally, let s and t be bisimilar process terms in ACP with linear recursion. At the start of this proof it was shown that $s = \langle X_1 | E_1 \rangle$ and $t = \langle Y_1 | E_2 \rangle$

where E_1 and E_2 are linear recursive specifications. Soundness of the axioms yields $\langle X_1|E_1\rangle \leftrightarrow s \leftrightarrow t \leftrightarrow \langle Y_1|E_2\rangle$, which implies $\langle X_1|E_1\rangle = \langle Y_1|E_2\rangle$. So $s = \langle X_1|E_1\rangle = \langle Y_1|E_2\rangle = t$. $\square$

Note that in the proof of Theorem 4.4.1, the procedure to equate bisimilar process terms is not based entirely on a term rewriting analysis. In general, automated verification tools (see Section 6.4) can be used to reduce process terms to normal form using the axiomatisation of ACP. However, applications of RDP and RSP often require human insight. Therefore, verifications of protocols in process algebra in practice ask for an interplay between a verification tool and its user. In this scenario the tool performs routine work, such as applications of rewrite rules, and the user provides manual input of tactics that involve RDP and RSP.

4.5 Approximation Induction Principle

At the start of the previous section we mentioned that $\mathcal{E}_{\mathrm{ACP}} + \mathrm{RDP}, \mathrm{RSP}$ is not complete for ACP with guarded recursion modulo bisimulation equivalence. In particular, we gave two symmetric guarded recursive specifications of the bag over $\{0,1\}$, and claimed that they cannot be proved equal by means of $\mathcal{E}_{\mathrm{ACP}} + \mathrm{RDP}, \mathrm{RSP}$. In this section we present an *approximation induction principle* (AIP), introduced by Bergstra and Klop [44] (see also [19]), which can be used to try and equate bisimilar guarded recursive specifications. Intuitively, AIP says that if two process terms are bisimilar up to any finite depth, then they are bisimilar.

Let $\mathbb{N}$ denote the collection of natural numbers $\{0, 1, 2, \ldots\}$. In order to formalise the notion of "bisimilar up to any finite depth", we need auxiliary unary *projection* operators π_n for $n \in \mathbb{N}$. The process term $\pi_n(t)$ can execute all transitions of t up to depth n, which is expressed by the following transition rules for $n \in \mathbb{N}$:

$$\frac{x \xrightarrow{v} \sqrt{}}{\pi_{n+1}(x) \xrightarrow{v} \sqrt{}} \qquad \frac{x \xrightarrow{v} x'}{\pi_{n+1}(x) \xrightarrow{v} \pi_n(x')}$$

The subscript n of the projection operator works as a counter, which is decreased by one at every transition of the subject term. Note that process terms $\pi_0(t)$ do not display any behaviour, so that they are bisimilar with δ.

Exercise 4.5.1. Compute the process graphs that belong to the process terms $\pi_n(\langle X \mid X{=}aY, Y{=}bX\rangle)$, for $n \in \mathbb{N}$.

Theorem 4.5.1. *ACP with projection operators and guarded recursion is a conservative extension of ACP with guarded recursion.*

Proof. The sources of the transition rules for the projection operators contain the fresh function symbol π_n. Since furthermore the transition rules of ACP with guarded recursion are source-dependent, the extension of this algebra with projection operators is conservative; see Theorem B.5.1. □

Theorem 4.5.2. *Bisimulation equivalence is a congruence with respect to ACP with projection operators and guarded recursion.*

Proof. This theorem follows from the fact that the transition rules for the projection operators, as well as of ACP with guarded recursion, are all in panth format; see Theorem B.3.1. □

Table 4.2 presents axioms for the projection operators, modulo bisimulation equivalence. Furthermore, Table 4.3 presents AIP, stating that two process terms are equal if all their projections are equal. The variables x and y in the axioms range over process terms, v ranges over A, and n ranges over $\mathbb{N}$.

Table 4.2. Axioms for projection operators

PR1	$\pi_n(x + y) = \pi_n(x) + \pi_n(y)$
PR2	$\pi_{n+1}(v) = v$
PR3	$\pi_{n+1}(v \cdot x) = v \cdot \pi_n(x)$
PR4	$\pi_0(x) = \delta$
PR5	$\pi_n(\delta) = \delta$

Table 4.3. Approximation induction principle

AIP	If $\pi_n(x) = \pi_n(y)$ for $n \in \mathbb{N}$, then $x = y$

Soundness of AIP for ACP with projection operators and guarded recursion modulo bisimulation equivalence was proved by van Glabbeek [98], using in an essential way the fact that the LTS generated by this algebra is *finitely branching* (see Definition B.1.1), meaning that each closed term has only finitely many initial transitions.

Theorem 4.5.3. $\mathcal{E}_{\text{ACP}} + \text{PR1-5} + \text{RDP}, \text{RSP}, \text{AIP}$ *is sound for ACP with projection operators and guarded recursion modulo bisimulation equivalence.*

Proof. Since bisimulation is both an equivalence and a congruence, we only need to check that if $s = t$ is an axiom in $\mathcal{E}_{\text{ACP}} + \text{RDP}, \text{RSP}, \text{AIP}$ and σ a closed substitution that maps the variables in s and t to process terms, then $\sigma(s) \leftrightarrow \sigma(t)$. Soundness of RDP, RSP, and the axioms in $\mathcal{E}_{\text{ACP}}$ can be checked as before. Here, we only provide some intuition for soundness of the axioms in Table 4.2:

- PR1 says that $\pi_n(s+t)$ can execute transitions of s and t up to depth n;
- PR2 says that $\pi_{n+1}(a)$ executes action a to terminate successfully;
- PR3 says that $\pi_{n+1}(at)$ executes action a, after which it executes transitions of t up to depth n;
- PR4,5 say that $\pi_0(t)$ and $\pi_n(\delta)$ do not execute any transitions.

These intuitions can be made rigorous by means of explicit bisimulation relations between the left- and right-hand sides of closed instantiations of the axioms in Table 4.2.

We proceed with a detailed proof of the soundness of AIP. Let s_0 and t_0 be process terms with $\pi_n(s_0) \leftrightarrow \pi_n(t_0)$ for $n \in \mathbb{N}$. We want to to find a bisimulation relation B that relates s_0 and t_0. We define that $s\,B\,t$ if and only if $\pi_n(s) \leftrightarrow \pi_n(t)$ for $n \in \mathbb{N}$. Clearly $s_0\,B\,t_0$; we proceed to show that B is a bisimulation relation.

Let $s\,B\,t$ and $s \xrightarrow{a} \sqrt{}$. Then $\pi_1(s) \xrightarrow{a} \sqrt{}$, so $\pi_1(s) \leftrightarrow \pi_1(t)$ yields $\pi_1(t) \xrightarrow{a} \sqrt{}$. Thus, $t \xrightarrow{a} \sqrt{}$. Likewise, $t \xrightarrow{a} \sqrt{}$ implies $s \xrightarrow{a} \sqrt{}$.

Let $s\,B\,t$ and $s \xrightarrow{a} s'$. We define the following sets of process terms for $n \in \mathbb{N}$:

$$S_n \triangleq \{t' \mid t \xrightarrow{a} t' \text{ and } \pi_n(s') \leftrightarrow \pi_n(t')\}.$$

We make three observations on the sets S_n for $n \in \mathbb{N}$.

1. Since $\pi_{n+1}(s) \leftrightarrow \pi_{n+1}(t)$ and $\pi_{n+1}(s) \xrightarrow{a} \pi_n(s')$, there exists a t' with $t \xrightarrow{a} t'$ and $\pi_n(s') \leftrightarrow \pi_n(t')$; hence, S_n is not empty.
2. There are only finitely many process terms t' such that $t \xrightarrow{a} t'$ (see Exercise 4.5.2), so S_n is finite.
3. Since $\pi_{n+1}(s') \leftrightarrow \pi_{n+1}(t')$ implies $\pi_n(s') \leftrightarrow \pi_n(t')$ (see Exercise 4.5.3), we have $S_n \supseteq S_{n+1}$.

These three observations together imply that the sets S_n for $n \in \mathbb{N}$ have a non-empty intersection. Select a process term t' in this intersection. Then $t \xrightarrow{a} t'$, and $\pi_n(s') \leftrightarrow \pi_n(t')$ for all $n \in \mathbb{N}$, so by the definition of B we have $s'\,B\,t'$. Likewise we can show that $s\,B\,t$ and $t \xrightarrow{a} t'$ implies $s \xrightarrow{a} s'$ with $s'\,B\,t'$.

Hence, B is a bisimulation relation, and so $s_0 \leftrightarrow t_0$. $\square$

Exercise 4.5.2. Prove that the LTS generated by ACP with projection operators and guarded recursion is finitely branching.

Exercise 4.5.3. Prove that $\pi_{n+1}(s) \leftrightarrow \pi_{n+1}(t)$ implies $\pi_n(s) \leftrightarrow \pi_n(t)$ for all process terms s and t.

Exercise 4.5.4. Give non-empty (not necessarily finite) sets S_n for $n \in \mathbb{N}$ such that $S_0 \supseteq S_1 \supseteq S_2 \supseteq \cdots$ and the intersection of all these sets is empty.

Exercise 4.5.5. Give two non-bisimilar, infinitely branching process graphs that are bisimilar up to any finite depth.

The following theorem bears witness to the strength of AIP. Note that RSP is not needed to derive the equations $\pi_n(s) = \pi_n(t)$.

Theorem 4.5.4. *For each pair of bisimilar process terms s and t in ACP with projection operators and guarded recursion, the equations $\pi_n(s) = \pi_n(t)$ for $n \in \mathbb{N}$ can be derived from $\mathcal{E}_{ACP} + PR1\text{-}5 + RDP$.*

Proof. Each process term t in ACP with projection operators and guarded recursion can be equated to a process term of the form

$$a_1 t_1 + \cdots + a_k t_k + b_1 + \cdots + b_\ell$$

by means of $\mathcal{E}_{ACP} + RDP$ (cf. Exercise 4.3.6).

Let $s \underleftrightarrow{} t$, and fix an $n \in \mathbb{N}$. By the observation above, $\pi_n(s)$ and $\pi_n(t)$ can be equated to process terms s' and t', respectively, in BPA extended with δ. Moreover, since bisimulation equivalence is a congruence, $s \underleftrightarrow{} t$ implies $\pi_n(s) \underleftrightarrow{} \pi_n(t)$. So soundness of the axioms yields $s' \underleftrightarrow{} \pi_n(s) \underleftrightarrow{} \pi_n(t) \underleftrightarrow{} t'$. Then, by completeness of the axiomatisation of BPA extended with δ modulo bisimulation equivalence, $s' = t'$. Hence, $\pi_n(s) = s' = t' = \pi_n(t)$. $\square$

Given two bisimilar process terms in ACP with projection operators and guarded recursion, Theorem 4.5.4 implies that all their projections are provably equal. So by AIP the two process terms themselves are provably equal. However, assuming it is unknown that the two process terms are bisimilar, one cannot derive equality of their projections one by one, as there are infinitely many such projections. Hence, some inductive argument is needed to master these derivations.

Example 4.5.1. We equate the following two symmetric guarded recursive specifications E and E' of the bag over $\{0, 1\}$:

$$X = in(0) \cdot (X \| out(0)) + in(1) \cdot (X \| out(1))$$

$$Y = in(0) \cdot (out(0) \| Y) + in(1) \cdot (out(1) \| Y).$$

This derivation is based on an application of AIP. First, we prove by induction on $n \in \mathbb{N}$ that

$$\pi_n((\cdots (((\langle X | E \rangle \| out(d_1)) \| out(d_2)) \| \cdots) \| out(d_k))$$
$$= \pi_n(out(d_k) \| (\cdots \| (out(d_2) \| (out(d_1) \| \langle Y | E' \rangle))) \cdots)) \tag{4.1}$$

for sequences $d_1 \cdots d_k$ of elements in $\{0, 1\}$. The base case $n \equiv 0$ is trivial, because then both process terms can be equated to δ by an application of

PR4. We focus on the inductive case, assuming that (4.1) has already been proved for $n \in \{1, \ldots, m\}$. For finite data sets Δ, let $\sum_{d \in \Delta} t(d)$ denote the alternative composition of process terms $t(d)$ for all elements d in Δ. (For example, if Δ is $\{0, 1\}$, then it denotes $t(0) + t(1)$.) Using RDP, $\mathcal{E}_{\text{ACP}}$, and induction we derive:

$$\pi_{m+1}((\cdots(\langle X|E\rangle \| out(d_1)) \| \cdots) \| out(d_k))$$

$$= \sum_{d \in \{0,1\}} in(d) \cdot \pi_m((\cdots(((\langle X|E\rangle \| out(d)) \| out(d_1)) \| \cdots) \| out(d_k))$$

$$+ \sum_{i \in \{1,\ldots,k\}} out(d_i) \cdot \pi_m((\cdots(\langle X|E\rangle \| \cdots \| out(d_{i-1})) \| out(d_{i+1})) \| \cdots)$$

$$= \sum_{d \in \{0,1\}} in(d) \cdot \pi_m(out(d_k) \| (\cdots \| (out(d_1) \| (out(d) \| \langle Y|E'\rangle)) \cdots))$$

$$+ \sum_{i \in \{1,\ldots,k\}} out(d_i) \cdot \pi_m(\cdots \| (out(d_{i+1}) \| (out(d_{i-1}) \| \cdots \| \langle Y|E'\rangle) \cdots))$$

$$= \pi_{m+1}(out(d_k) \| (\cdots \| (out(d_1) \| \langle Y|E'\rangle)) \cdots)).$$

This concludes the derivation of (4.1) for $n \in \mathbb{N}$. By AIP it follows that

$$(\cdots(((\langle X|E\rangle \| out(d_1)) \| out(d_2)) \| \cdots) \| out(d_k)$$

$$= out(d_k) \| (\cdots \| (out(d_2) \| (out(d_1) \| \langle Y|E'\rangle)) \cdots).$$

In particular, the case $k \equiv 0$ yields the desired equation $\langle X|E\rangle = \langle Y|E'\rangle$.

Exercise 4.5.6. Derive $\langle X \mid X{=}aXb{+}b \rangle = \langle Y \mid Y{=}aZb{+}b, Z{=}aYb{+}b \rangle$ from $\mathcal{E}_{\text{ACP}} + $ PR1-5 $+$ RDP, AIP.

5. Abstraction

If a customer asks a programmer to implement a product, ideally this customer is able to provide the external behaviour of the desired program. That is, he or she is able to tell what should be the output of the program for each possible input. The programmer then comes up with an implementation. The question is, does this implementation really display the desired external behaviour? To answer this question, we need to abstract away from the internal computation steps of the program.

5.1 Rooted Branching Bisimulation Equivalence

In order to abstract away from internal actions, Milner [151] introduced a special constant τ, called the *silent step*. Intuitively, a τ-transition represents a sequence of internal actions that can be eliminated from a process graph. As any atomic action, the constant τ can execute itself, after which it terminates successfully. This is expressed by the transition rule

$$\frac{}{\tau \xrightarrow{\tau} \sqrt{}}$$

From now on, v and w in the transition rules and the axioms of ACP with guarded recursion range over $A \cup \{\tau\}$. (So the transition rule for atomic actions in Table 2.1 yields the transition rule for the silent step τ presented above.) The domain of the communication function γ is extended with the silent step, $\gamma : A \cup \{\tau\} \times A \cup \{\tau\} \to A \cup \{\delta\}$, by defining that each communication involving τ results to δ.

In the presence of the silent step τ, bisimulation is no longer a satisfactory process equivalence. Namely, if processes p and q are equivalent, and p can execute an action τ, then it need not be the case that q can simulate this τ-transition of p by the execution of an action τ. The intuition for the silent step, that it represents an internal computation in which we are not really interested, asks for a new process equivalence. The question that we must pose ourselves is:

which τ-transitions are truly silent?

The obvious answer to this question, "all τ-transitions are truly silent", turns out to be incorrect. Namely, this answer would produce an equivalence relation that is not a congruence.

As an example of an action τ that is not truly silent, consider the process terms $a + \tau\delta$ and a. If the τ in the first term were truly silent, then these two terms would be equivalent. However, the process graph of the first term contains a deadlock, $a + \tau\delta \xrightarrow{\tau} \delta$, while the process graph of the second term does not. Hence, the τ in the first term is not truly silent. In order to describe this case more vividly, we give an example.

Example 5.1.1. Consider a protocol that first receives a datum d via channel 1, and then communicates this datum via channel 2 or via channel 3. If the datum is communicated through channel 2, then it is sent into channel 4. If the datum is communicated through channel 3, then it gets stuck, as the subsequent channel 5 is broken. So the system gets into a deadlock if the datum d is transferred via channel 3. This deadlock should not disappear if we abstract away from the internal communication actions via channels 2 and 3, because this would cover up an important problem of the protocol.

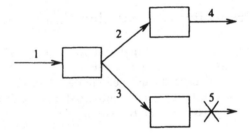

Fig. 5.1. Protocol with a malfunctioning channel

The system, which is depicted in Fig. 5.1, is described by the process term

$$\partial_{\{s_5(d)\}}(r_1(d) \cdot (c_2(d) \cdot s_4(d) + c_3(d) \cdot s_5(d)))$$
$$\overset{\text{D1,2,4,5}}{=} r_1(d) \cdot (c_2(d) \cdot s_4(d) + c_3(d) \cdot \delta)$$

where $s_i(d)$, $r_i(d)$, and $c_i(d)$ represent a send, read, and communication action of the datum d via channel i, respectively. Abstracting away from the internal actions $c_2(d)$ and $c_3(d)$ in this process term yields $r_1(d) \cdot (\tau \cdot s_4(d) + \tau \cdot \delta)$. The second τ in this term cannot be deleted, because then the process would no longer be able to get into a deadlock. Hence, this τ is not truly silent.

As a further example of a τ-transition that is not truly silent, consider the process terms $a + \tau b$ and $a + b$. We argued previously that the process terms $\partial_{\{b\}}(a + \tau b) = a + \tau\delta$ and $\partial_{\{b\}}(a + b) = a$ are not equivalent, because the first term contains a deadlock while the second term does not. Hence, $a + \tau b$ and $a + b$ cannot be equivalent, for else the envisioned equivalence relation would not be a congruence.

Problems with congruence can be avoided by taking a more restrictive view on abstracting away from silent steps. A correct answer to the question

which τ-transitions are truly silent?

turns out to be

those τ-transitions that do not lose possible behaviours !

For example, the process terms $a + \tau(a + b)$ and $a + b$ are equivalent, because the τ in the first process term is truly silent: after execution of this τ it is still possible to execute a. In general, process terms $s + \tau(s + t)$ and $s + t$ are equivalent for all process terms s and t. By contrast, in a process term such as $a + \tau b$ the τ is not truly silent, since execution of this τ means losing the option to execute a.

The intuition above is formalised in the notion of *branching bisimulation equivalence* (see Definition B.4.1). Let the processes p and q be branching bisimilar. If $p \xrightarrow{\tau} p'$, then q does not have to simulate this τ-transition if it is truly silent, meaning that p' and q are branching bisimilar. Moreover, a non-silent transition $p \xrightarrow{a} p'$ need not be simulated by q immediately, but only after a number of truly silent τ-transitions: $q \xrightarrow{\tau} \cdots \xrightarrow{\tau} q_0 \xrightarrow{a} q'$, where p and q_0 are branching bisimilar (to ensure that the τ-transitions are truly silent) and p' and q' are branching bisimilar (so that $p \xrightarrow{a} p'$ is simulated by $q_0 \xrightarrow{a} q'$). A special termination predicate $\downarrow$ is needed in order to relate branching bisimilar process terms such as $a\tau$ and a. Definition B.4.1 is presented below for the relations $\xrightarrow{a}$ for $a \in A$ and the predicate $\downarrow$.

Definition 5.1.1 (Branching bisimulation). *Assume a special termination predicate $\downarrow$, and let $\sqrt{}$ represent a state with $\sqrt{} \downarrow$. A branching bisimulation relation $\mathcal{B}$ is a binary relation on the collection of processes such that:*

1. *if $p\,\mathcal{B}\,q$ and $p \xrightarrow{a} p'$, then*
   - *either $a \equiv \tau$ and $p'\,\mathcal{B}\,q$;*
   - *or there is a sequence of (zero or more) τ-transitions $q \xrightarrow{\tau} \cdots \xrightarrow{\tau} q_0$ such that $p\,\mathcal{B}\,q_0$ and $q_0 \xrightarrow{a} q'$ with $p'\,\mathcal{B}\,q'$;*
2. *if $p\,\mathcal{B}\,q$ and $q \xrightarrow{a} q'$, then*
   - *either $a \equiv \tau$ and $p\,\mathcal{B}\,q'$;*
   - *or there is a sequence of (zero or more) τ-transitions $p \xrightarrow{\tau} \cdots \xrightarrow{\tau} p_0$ such that $p_0\,\mathcal{B}\,q$ and $p_0 \xrightarrow{a} p'$ with $p'\,\mathcal{B}\,q'$.*
3. *if $p\,\mathcal{B}\,q$ and $p \downarrow$, then there is a sequence of (zero or more) τ-transitions $q \xrightarrow{\tau} \cdots \xrightarrow{\tau} q_0$ such that $q_0 \downarrow$;*
4. *if $p\,\mathcal{B}\,q$ and $q \downarrow$, then there is a sequence of (zero or more) τ-transitions $p \xrightarrow{\tau} \cdots \xrightarrow{\tau} p_0$ such that $p_0 \downarrow$.*

Two processes p and q are branching bisimilar, *denoted by $p \leftrightarrow_b q$, if there is a branching bisimulation relation $\mathcal{B}$ such that $p \mathcal{B} q$.*

Example 5.1.2. $a + \tau(a + b) \leftrightarrow_b \tau(a + b) + b$.
A branching bisimulation relation that relates these two process terms is defined by $a + \tau(a + b) \mathcal{B} \tau(a + b) + b$, $a + b \mathcal{B} \tau(a + b) + b$, $a + \tau(a + b) \mathcal{B} a + b$, $a + b \mathcal{B} a + b$, and $\sqrt{} \mathcal{B} \sqrt{}$. This relation can be depicted as follows:

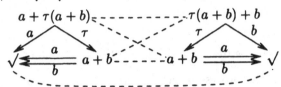

It is left to the reader to verify that this relation satisfies the requirements of a branching bisimulation.

Exercise 5.1.1. Give branching bisimulation relations to prove that the process terms a, $a\tau$, and τa are branching bisimilar.

Exercise 5.1.2. Give a branching bisimulation relation to prove that the process terms $\tau(\tau(a + b) + b) + a$ and $a + b$ are branching bisimilar.

Exercise 5.1.3. Explain why $\tau a + \tau b$ and $a + b$ are not branching bisimilar.

Branching bisimilarity is an equivalence relation; see [32]. Branching bisimulation equivalence, however, is not a congruence with respect to BPA. For example, τa and a are branching bisimilar (see Exercise 5.1.1), but $\tau a + b$ and $a + b$ are not branching bisimilar. Namely, if $\tau a + b$ executes τ then it loses the option to execute b, so this τ-transition is not truly silent.

Milner [155] showed that this problem can be overcome by adding a rootedness condition: initial τ-transitions are never truly silent. In other words, two processes are considered equivalent if they can simulate each other's initial transitions, such that the resulting processes are branching bisimilar. This leads to the notion of *rooted branching bisimulation equivalence*; see Definition B.4.2. This definition is presented below for the relations $\xrightarrow{a} \sqrt{}$ for $a \in A$ and the predicate $\downarrow$.

Definition 5.1.2 (Rooted branching bisimulation). *Assume the termination predicate $\downarrow$, and let $\sqrt{}$ represent a state with $\sqrt{} \downarrow$. A* rooted branching bisimulation relation $\mathcal{B}$ *is a binary relation on processes such that:*

1. *if $p \mathcal{B} q$ and $p \xrightarrow{a} p'$, then $q \xrightarrow{a} q'$ with $p' \leftrightarrow_b q'$;*
2. *if $p \mathcal{B} q$ and $q \xrightarrow{a} q'$, then $p \xrightarrow{a} p'$ with $p' \leftrightarrow_b q'$;*
3. *if $p \mathcal{B} q$ and $p \downarrow$, then $q \downarrow$;*
4. *if $p \mathcal{B} q$ and $q \downarrow$, then $p \downarrow$.*

Two processes p and q are rooted branching bisimilar, *denoted by $p \leftrightarrow_{rb} q$, if there is a rooted branching bisimulation relation $\mathcal{B}$ such that $p \mathcal{B} q$.*

Since branching bisimilarity is an equivalence relation, it is not hard to see that rooted branching bisimilarity is also an equivalence relation. Branching bisimulation equivalence strictly includes rooted branching bisimulation equivalence, which in turn strictly includes bisimulation equivalence:

$$\underline{\leftrightarrow} \subset \underline{\leftrightarrow}_{rb} \subset \underline{\leftrightarrow}_b \ .$$

In the absence of τ (for example, in ACP), bisimulation and branching bisimulation induce exactly the same equivalence classes. In other words, two process terms in ACP are bisimilar if and only if they are branching bisimilar.

Exercise 5.1.4. Say for the following five pairs of process terms whether or not they are bisimilar, rooted branching bisimilar, or branching bisimilar:

- $(a + b)(c + d)$ and $ac + ad + bc + bd$;
- $(a + b)(c + d)$ and $(b + a)(d + c) + a(c + d)$;
- $\tau(b + a) + \tau(a + b)$ and $a + b$;
- $c(\tau(b + a) + \tau(a + b))$ and $c(a + b)$;
- $a(\tau b + c)$ and $a(b + \tau c)$.

In each case, give explicit relations, or explain why such relations do not exist.

Exercise 5.1.5. Prove that $a(s\|(\tau t)) \underline{\leftrightarrow}_{rb} a(s\|t)$ for process terms s and t.

Exercise 5.1.6. Verify that rooted branching bisimilarity is an equivalence relation.

5.2 Guarded Linear Recursion Revisited

Assume a recursive specification E that consists of linear recursive equations $X_i = t_i(X_1, \ldots, X_n)$ for $i \in \{1, \ldots, n\}$. Since from now on we consider processes in the setting of rooted branching bisimulation equivalence, processes $p_1, \ldots, p_n$ are said to be a solution for E (with respect to rooted branching bisimulation equivalence) if $p_i \underline{\leftrightarrow}_{rb} t_i(p_1, \ldots, p_n)$ for $i \in \{1, \ldots, n\}$ (cf. Definition 4.1.2).

In the setting with the silent step, the notion of guardedness (cf. Definition 4.1.3), which aims to classify those recursive specifications that have a unique solution modulo the process equivalence under consideration, needs to be adapted. For example, all process terms τs are solutions for the recursive specification $X = \tau X$, because $\tau s \underline{\leftrightarrow}_{rb} \tau\tau s$ holds for all process terms s. Hence, we consider such a recursive specification to be unguarded. The notion of guardedness is extended to linear recursive specifications (see Definition 4.4.1) that involve silent steps by requiring the absence of τ-loops.

Definition 5.2.1 (Guarded linear recursive specification). *A recursive specification is* linear *if its recursive equations are of the form*

$$X = a_1 X_1 + \cdots + a_k X_k + b_1 + \cdots + b_\ell$$

with $a_1, \ldots, a_k, b_1, \ldots, b_\ell \in A \cup \{\tau\}$.

A linear recursive specification E is guarded *if there does not exist an infinite sequence of τ-transitions $\langle X|E \rangle \xrightarrow{\tau} \langle X'|E \rangle \xrightarrow{\tau} \langle X''|E \rangle \xrightarrow{\tau} \cdots$.*

The guarded linear recursive specifications are exactly the linear recursive specifications that have a unique solution, modulo rooted branching bisimulation equivalence.

Exercise 5.2.1. Show that $\{X{=}aY{+}\tau Y, Y{=}bX{+}\tau X\}$ is not guarded. Give two solutions for this linear recursive specification that are not rooted branching bisimilar.

Theorem 5.2.1. *ACP with silent step and guarded linear recursion is a conservative extension of ACP with linear recursion.*

Proof. This theorem follows from the following three facts.

1. The transition rules of ACP and for linear recursive specifications that do not include a τ are all source-dependent.
2. The source of the transition rule for the silent step is the fresh constant τ. The source of a transition rule for a guarded linear recursive specification E that includes a τ is the fresh constant $\langle X|E \rangle$.
3. Each transition rule for alternative composition, sequential composition, or guarded linear recursion that involves τ-transitions, such as

$$\frac{x \xrightarrow{\tau} x'}{x + y \xrightarrow{\tau} x'}$$

   includes a premise containing the fresh relation symbol $\xrightarrow{\tau}$ or predicate $\xrightarrow{\tau} \sqrt{}$, and a left-hand side of which all variables occur in the source of the transition rule.

Hence, Theorem B.5.1 says that ACP with silent step and guarded linear recursion is a conservative extension of ACP with linear recursion. □

Theorem 5.2.2. *Rooted branching bisimulation equivalence is a congruence with respect to ACP with silent step and guarded linear recursion.*

Proof. The TSS of ACP with silent step and guarded linear recursion can be brought into RBB cool format (see Definition B.4.5), by incorporating the successful termination predicate $\downarrow$ from Definition 5.1.1 in the transition rules. That is, the symbol $\sqrt{}$ is added to the signature as a special constant, and the transition rule $\frac{}{\sqrt{}\downarrow}$ is added to the TSS. Moreover, transition rules that contain an occurrence of a predicate symbol $\xrightarrow{a} \sqrt{}$ are adapted accordingly.

It is left to the reader to verify that the adapted TSS of ACP with silent step and guarded linear recursion is RBB cool (see Exercise 5.2.2). This fact implies that the rooted branching bisimulation equivalence that this TSS induces is a congruence; see Theorem B.4.1. □

Exercise 5.2.2. Spell out the adapted TSS of ACP with silent step and guarded linear recursion from the proof of Theorem 5.2.2, and verify that it is RBB cool.

5.3 Axioms for the Silent Step

Table 5.1 presents the axioms B1,2 for the silent step, modulo rooted branching bisimulation equivalence. The variables x and y in the axioms range over process terms, while v ranges over $A \cup \{\tau\}$.

Table 5.1. Axioms for the silent step

B1	$v \cdot \tau = v$
B2	$v \cdot (\tau \cdot (x + y) + x) = v \cdot (x + y)$

Theorem 5.3.1. $\mathcal{E}_{ACP} + B1,2 + RDP, RSP$ *is sound for ACP with silent step and guarded linear recursion, modulo rooted branching bisimulation equivalence.*

Proof. Since rooted branching bisimulation is both an equivalence and a congruence, we only need to check that if $s = t$ is an axiom in $\mathcal{E}_{ACP} + B1,2 + RDP, RSP$ and σ a closed substitution that maps the variables in s and t to process terms, then $\sigma(s) \underline{\leftrightarrow}_{rb} \sigma(t)$. Soundness of the axioms in $\mathcal{E}_{ACP} + RDP, RSP$ can be checked as before.

The axioms B1,2 say that a non-initial τ-transition that does not lose any possible behaviour is truly silent. This intuition can be made rigorous by means of explicit rooted branching bisimulation relations between the left- and right-hand sides of closed instantiations of B1,2. □

Exercise 5.3.1. Prove that each process term in ACP with silent step and guarded linear recursion generates a regular process graph that does not contain a loop of τ-transitions.

Exercise 5.3.2. Suppose we would allow $\gamma(a, b) \equiv \tau$. Give an example of two guarded recursive specifications of which the merge can only be described by an unguarded recursive specification.

The following completeness result is due to van Glabbeek [101]. (See [45, 156] for similar completeness results with respect to observation equivalence.)

Theorem 5.3.2. $\mathcal{E}_{\mathrm{ACP}} + \mathrm{B}1,2 + \mathrm{RDP}, \mathrm{RSP}$ *is complete for ACP with silent step and guarded linear recursion, modulo rooted branching bisimulation equivalence.*

Proof. As a first step we note that each process term t_1 in ACP with silent step and guarded linear recursion is provably equal to a process term $\langle X_1 | E \rangle$ with E a guarded linear recursive specification. Namely, each such process term t_1 generates a regular process graph that does not contain a loop of τ-transitions (see Exercise 5.3.1), with states say $t_1, \ldots, t_n$. (This observation uses in an essential way that communications between atomic actions do not result to τ; see Exercise 5.3.2.) This process graph can be expressed in the form of equations

$$t_i = a_{i1} t_{i_1} + \cdots + a_{ik_i} t_{ik_i} + b_{i1} + \cdots + b_{il_i}$$

for $i \in \{1, \ldots, n\}$ (cf. Exercise 4.3.6). Let the guarded linear recursive specification E consist of the recursive equations

$$X_i = a_{i1} X_{i_1} + \cdots + a_{ik_i} X_{ik_i} + b_{i1} + \cdots + b_{il_i}$$

for $i \in \{1, \ldots, n\}$. Since replacing X_i by t_i for $i \in \{1, \ldots, n\}$ is a solution for E, RSP yields $t_1 = \langle X_1 | E \rangle$.

It remains to prove that if $\langle X_1 | E_1 \rangle \underleftrightarrow{}_{rb} \langle Y_1 | E_2 \rangle$ for guarded linear recursive specifications E_1 and E_2, then $\langle X_1 | E_1 \rangle = \langle Y_1 | E_2 \rangle$. First, suppose E_1 contains a recursive equation $W = \tau + \cdots + \tau$ with $W \not\equiv X_1$. Let E_1' be obtained from E_1 by removing the recursive equation for W, and replacing summands aW in right-hand sides of recursive equations of E_1 by a. Using RDP, A3, and B1, it can be derived that substituting process terms $\langle X | E_1 \rangle$ for recursion variables X in E_1' is a solution for E_1'. So by RSP we have $\langle X_1 | E_1 \rangle = \langle X_1 | E_1' \rangle$. Thus, recursive equations $W = \tau + \cdots + \tau$ with $W \not\equiv X_1$ can be eliminated from E_1, and likewise such recursive equations with $W \not\equiv Y_1$ can be eliminated from E_2.

Let E_1 and E_2 consist of linear recursive equations $X = t_X$ for $X \in \mathcal{X}$ and $Y = t_Y$ for $Y \in \mathcal{Y}$, respectively, where the t_X for $X \not\equiv X_1$ and the t_Y for $Y \not\equiv Y_1$ are not of the form $\tau + \cdots + \tau$. The recursive specification E is defined to consist of the linear recursive equations $Z_{XY} = t_{XY}$ for $X \in \mathcal{X}$ and $Y \in \mathcal{Y}$ with $\langle X | E_1 \rangle \underleftrightarrow{}_b \langle Y | E_2 \rangle$, where t_{XY} consists of the following summands:

1. t_{XY} contains a summand $aZ_{X'Y'}$ if and only if t_X and t_Y contain the summands aX' and aY', respectively, and $\langle X' | E_1 \rangle \underleftrightarrow{}_b \langle Y' | E_2 \rangle$;
2. t_{XY} contains a summand b if and only if both t_X and t_Y contain the summand b;
3. t_{XY} contains a summand $\tau Z_{X'Y}$ if and only if $XY \not\equiv X_1 Y_1$, t_X contains the summand $\tau X'$, and $\langle X' | E_1 \rangle \underleftrightarrow{}_b \langle Y | E_2 \rangle$;

4. t_{XY} contains a summand $\tau Z_{XY'}$ if and only if $XY \not\equiv X_1Y_1$, t_Y contains the summand $\tau Y'$, and $\langle X|E_1\rangle \underline{\leftrightarrow}_b \langle Y'|E_2\rangle$.

Since E_1 and E_2 are guarded, all chains $\langle X|E_1\rangle \overset{\tau}{\to} \langle X'|E_1\rangle \overset{\tau}{\to} \langle X''|E_1\rangle \overset{\tau}{\to} \cdots$ and $\langle Y|E_2\rangle \overset{\tau}{\to} \langle Y'|E_2\rangle \overset{\tau}{\to} \langle Y''|E_2\rangle \overset{\tau}{\to} \cdots$ are finite (see Definition 5.2.1). This implies that all chains $\langle Z_{XY}|E\rangle \overset{\tau}{\to} \langle Z_{X'Y'}|E\rangle \overset{\tau}{\to} \langle Z_{X''Y''}|E\rangle \overset{\tau}{\to} \cdots$ are finite, so E is guarded.

For recursion variables Z_{XY} in E, let the process term u_{XY} consist of the following summands:

1. u_{XY} contains a summand $a\langle X'|E_1\rangle$ if and only if t_X and t_Y contain summands aX' and aY', respectively, with $\langle X'|E_1\rangle \underline{\leftrightarrow}_b \langle Y'|E_2\rangle$;
2. u_{XY} contains a summand b if and only if both t_X and t_Y contain the summand b;
3. u_{XY} contains a summand $\tau\langle X'|E_1\rangle$ if and only if $XY \not\equiv X_1Y_1$, t_X contains the summand $\tau X'$, and $\langle X'|E_1\rangle \underline{\leftrightarrow}_b \langle Y|E_2\rangle$.

Furthermore, for recursion variables Z_{XY} in E, let the process term s_{XY} be defined by:

$$s_{XY} \triangleq \begin{cases} \tau\langle X|E_1\rangle + u_{XY} & \text{if } XY \not\equiv X_1Y_1 \text{ and } t_Y \text{ contains a summand} \\ & \tau Y' \text{ with } \langle X|E_1\rangle \underline{\leftrightarrow}_b \langle Y'|E_2\rangle, \\ \langle X|E_1\rangle & \text{otherwise.} \end{cases}$$

By RDP and A3,

$$\langle X|E_1\rangle = \langle X|E_1\rangle + u_{XY}. \tag{5.1}$$

So for $a \in A \cup \{\tau\}$,

$$\begin{aligned} a(\tau\langle X|E_1\rangle + u_{XY}) &\overset{(5.1)}{=} a(\tau(\langle X|E_1\rangle + u_{XY}) + u_{XY}) \\ &\overset{B2}{=} a(\langle X|E_1\rangle + u_{XY}) \\ &\overset{(5.1)}{=} a\langle X|E_1\rangle. \end{aligned}$$

Hence, for $a \in A \cup \{\tau\}$ and recursion variables Z_{XY} in E, the definition of s_{XY} yields

$$as_{XY} = a\langle X|E_1\rangle. \tag{5.2}$$

Let the substitutions σ and ψ from recursion variables to process terms be defined as follows:

- σ maps recursion variables X in E_1 to $\langle X|E_1\rangle$;
- ψ maps recursion variables Z_{XY} in E to s_{XY}.

We proceed to show that substituting s_{XY} for recursion variables Z_{XY} in E is a solution for E; that is, $s_{XY} = \psi(t_{XY})$ for recursion variables Z_{XY} in E. We distinguish two cases, depending on whether or not $XY \equiv X_1Y_1$.

1. Let $XY \equiv X_1Y_1$.

   By assumption, $\langle X_1|E_1 \rangle \mathbin{\underline{\leftrightarrow}}_{rb} \langle Y_1|E_2 \rangle$. Furthermore, E_1 and E_2 do not contain recursive equations $W = \tau + \cdots + \tau$ for recursion variables W unequal to X_1 and Y_1, respectively. These observations together imply that for each summand aX' of t_{X_1} there is a summand aY' of t_{Y_1} with $\langle X'|E_1 \rangle \mathbin{\underline{\leftrightarrow}}_{b} \langle Y'|E_2 \rangle$. Likewise, each summand b of t_{X_1} is also a summand of t_{Y_1}. Hence, by the definition of $t_{X_1Y_1}$, for each summand aX' or b of t_{X_1} there is a summand $aZ_{X'Y'}$ or b of $t_{X_1Y_1}$. Vice versa, by the definition of $t_{X_1Y_1}$, for each summand $aZ_{X'Y'}$ or b of $t_{X_1Y_1}$ there is a summand aX' or b of t_{X_1}. Hence, each summand $a\langle X'|E_1 \rangle$ or b of $\sigma(t_{X_1})$ corresponds to a summand $as_{X'Y'}$ or b of $\psi(t_{X_1Y_1})$, and vice versa. So by equation (5.2) together with A3, $\sigma(t_{X_1}) = \psi(t_{X_1Y_1})$. Hence,

$$s_{X_1Y_1} \equiv \langle X_1|E_1 \rangle \stackrel{\text{RDP}}{=} \sigma(t_{X_1}) = \psi(t_{X_1Y_1}). \tag{5.3}$$

2. Let $XY \not\equiv X_1Y_1$.

   Once more we distinguish two cases.

2.1. Let t_Y not contain a summand $\tau Y'$ with $\langle X|E_1 \rangle \mathbin{\underline{\leftrightarrow}}_{b} \langle Y'|E_2 \rangle$.

   Z_{XY} is a recursion variable in E, so $\langle X|E_1 \rangle \mathbin{\underline{\leftrightarrow}}_{b} \langle Y|E_2 \rangle$. By assumption, t_Y does not contain a summand $\tau Y'$ with $\langle X|E_1 \rangle \mathbin{\underline{\leftrightarrow}}_{b} \langle Y'|E_2 \rangle$. Furthermore, E_1 and E_2 do not contain recursive equations $W = \tau + \cdots + \tau$ for recursion variables W unequal to X_1 and Y_1. These observations together imply that for each summand aX' of t_X with a not a truly silent τ (i.e., $a \not\equiv \tau$ or $\langle X'|E_1 \rangle \mathbin{\not\underline{\leftrightarrow}}_{b} \langle Y|E_2 \rangle$), there is a summand aY' of t_Y with $\langle X'|E_1 \rangle \mathbin{\underline{\leftrightarrow}}_{b} \langle Y'|E_2 \rangle$. Likewise, each summand b of t_X is also a summand of t_Y. Hence, the first three clauses in the definition of t_{XY} yield that for each summand aX' or b of t_X there is a summand $aZ_{X'Y'}$ or b of t_{XY}. Since t_Y does not contain a summand $\tau Y'$ with $\langle X|E_1 \rangle \mathbin{\underline{\leftrightarrow}}_{b} \langle Y'|E_2 \rangle$, the fourth clause in the definition of t_{XY} is vacuous. So vice versa, the first three clauses in the definition of t_{XY} yield that for each summand $aZ_{X'Y'}$ or b of t_{XY} there is a summand aX' or b of t_X. Hence, each summand $a\langle X'|E_1 \rangle$ or b of $\sigma(t_X)$ corresponds to a summand $as_{X'Y'}$ or b of $\psi(t_{XY})$, and vice versa. So by equation (5.2) together with A3, $\sigma(t_X) = \psi(t_{XY})$. Hence,

$$s_{XY} \equiv \langle X|E_1 \rangle \stackrel{\text{RDP}}{=} \sigma(t_X) = \psi(t_{XY}). \tag{5.4}$$

2.2. Let t_Y contain one or more summands $\tau Y'$ with $\langle X|E_1 \rangle \mathbin{\underline{\leftrightarrow}}_{b} \langle Y'|E_2 \rangle$.

   Then, by the fourth clause in its definition, t_{XY} contains one or more summands $\tau Z_{XY'}$. Furthermore, by the first three clauses in the definition of t_{XY} together with the definition of u_{XY}, for each remaining summand b or $aZ_{X'Y'}$ (with $a \not\equiv \tau$ or $X' \not\equiv X$) of t_{XY} there is a summand b or $a\langle X'|E_1 \rangle$ of u_{XY}, and vice versa. Hence, $\psi(t_{XY})$ contains one or more summands $\tau s_{XY'}$, while each remaining summand b or $as_{X'Y'}$ (with $a \not\equiv \tau$ or $X' \not\equiv X$) of $\psi(t_{XY})$ corresponds with a summand b or $a\langle X'|E_1 \rangle$ of u_{XY}, and vice versa. So

$$s_{XY} \equiv \tau\langle X|E_1\rangle + u_{XY} \overset{(5.2),\text{A3}}{=} \psi(t_{XY}). \tag{5.5}$$

We conclude from equations (5.3), (5.4), and (5.5) that substituting process terms s_{XY} for recursion variables Z_{XY} in E is a solution for E. Then RSP yields $s_{XY} = \langle Z_{XY}|E\rangle$ for recursion variables Z_{XY} in E, so in particular $\langle X_1|E_1\rangle = \langle Z_{X_1Y_1}|E\rangle$. Likewise we can derive $\langle Y_1|E_2\rangle = \langle Z_{X_1Y_1}|E\rangle$. So

$$\langle X_1|E_1\rangle = \langle Z_{X_1Y_1}|E\rangle = \langle Y_1|E_2\rangle.$$

Finally, let s and t be rooted branching bisimilar process terms in ACP with silent step and guarded linear recursion. At the start of this proof it was shown that $s = \langle X_1|E_1\rangle$ and $t = \langle Y_1|E_2\rangle$ where E_1 and E_2 are guarded linear recursive specifications. Soundness of the axioms yields $\langle X_1|E_1\rangle \underset{rb}{\leftrightarrow} s \underset{rb}{\leftrightarrow} t \underset{rb}{\leftrightarrow} \langle Y_1|E_2\rangle$, which implies $\langle X_1|E_1\rangle = \langle Y_1|E_2\rangle$. So $s = \langle X_1|E_1\rangle = \langle Y_1|E_2\rangle = t$. $\square$

Exercise 5.3.3. Derive the next equations from $\mathcal{E}_{\text{ACP}} + \text{B1}, 2 + \text{RDP}, \text{RSP}$:

- $a(\tau b + b) = ab$;
- $a(\tau(b + c) + b) = a(\tau(b + c) + c)$;
- $a(s\|(\tau t)) = a(s\|t)$ for process terms s and t;
- $\langle X \mid X{=}aY, Y{=}\tau X\rangle = \langle Z \mid Z{=}aZ\rangle$;
- $\langle X \mid X{=}(a{+}b)Y, Y{=}(\tau{+}b)X\rangle = \langle Z \mid Z{=}(a{+}b)Z\rangle$.

Exercise 5.3.4. Give a counter-example to show that rooted branching bisimulation equivalence is not a congruence in the presence of the projection operators from Section 4.5. Why is it not possible to bring the second transition rule of the projection operators into RBB cool format?

Exercise 5.3.5. Adapt the transition rules for the projection operators, so that τ-transitions do not decrease the counter n. Verify that the resulting transition rules can be brought into RBB cool format, by incorporating the successful termination predicate $\downarrow$. Give axioms for the adapted interpretation of projection operators.

Exercise 5.3.6. Prove soundness of AIP for ACP with silent step modulo rooted branching bisimulation equivalence, for the adapted interpretation of the projection operators from the previous exercise.

5.4 Abstraction Operators

We introduce unary *abstraction* operators τ_I, for subsets I of A, which rename all atomic actions in I into τ. The abstraction operators, which enable us to abstract away from the internal computation steps of an implementation, were introduced by Bergstra and Klop [43]. The behaviour of the abstraction operators is captured by the following transition rules, which express that in $\tau_I(t)$ all labels of transitions of t that are in I are renamed into τ:

$$\frac{x \overset{v}{\to} \sqrt{}}{\tau_I(x) \overset{v}{\to} \sqrt{}} \quad v \notin I \qquad\qquad \frac{x \overset{v}{\to} x'}{\tau_I(x) \overset{v}{\to} \tau_I(x')} \quad v \notin I$$

$$\frac{x \overset{v}{\to} \sqrt{}}{\tau_I(x) \overset{\tau}{\to} \sqrt{}} \quad v \in I \qquad\qquad \frac{x \overset{v}{\to} x'}{\tau_I(x) \overset{\tau}{\to} \tau_I(x')} \quad v \in I$$

The variables x and x' range over process terms, while v ranges over $A \cup \{\tau\}$. ACP extended with silent step and abstraction operators is denoted by ACP_τ.

Exercise 5.4.1. Let $\gamma(a,b) \overset{\Delta}{=} c$. Derive the transition

$$\tau_{\{c\}}(\partial_{\{a,b\}}((aa) \| (bb))) \overset{\tau}{\to} \tau_{\{c\}}(\partial_{\{a,b\}}(a \| b))$$

from the transition rules of ACP_τ.

Exercise 5.4.2. Show that the process term $\tau_{\{a\}}(\langle X \mid X = aX \rangle)$ and the deadlock δ are branching bisimilar.

Exercise 5.4.3. Prove that for each process term t in ACP_τ, the process term $\tau_A(t)$ is branching bisimilar to τ, $\tau\delta$, or $\tau + \tau\delta$.

Exercise 5.4.4. Give a counter-example to show that in general the equation $\tau_I(\partial_H(x)) = \partial_H(\tau_I(x))$ is not sound modulo rooted branching bisimulation equivalence.

Exercise 5.4.5. Let t_1 and t_2 be process terms with $t_1 \underline{\leftrightarrow}_{rb} at_2$ and $t_2 \underline{\leftrightarrow}_{rb} \tau t_2$. Can it be concluded from these two equivalences that t_1 is rooted branching bisimilar with $\tau_{\{b\}}(\langle X_1 \mid X_1 = aX_2, X_2 = bX_2 \rangle)$?

Theorem 5.4.1. *ACP_τ with guarded linear recursion is a conservative extension of ACP with silent step and guarded linear recursion.*

Proof. This theorem follows from the following two facts.

1. The transition rules of ACP, the silent step, and guarded linear recursion are all source-dependent.
2. The sources of the transition rules for the abstraction operators contain an occurrence of τ_I.

Hence, Theorem B.5.1 says that ACP_τ with guarded linear recursion is a conservative extension of ACP with silent step and guarded linear recursion. $\square$

Theorem 5.4.2. *Rooted branching bisimulation equivalence is a congruence with respect to ACP_τ with guarded linear recursion.*

Proof. As in the proof of Theorem 5.2.2, the transition rules of ACP_τ with guarded linear recursion can be brought into RBB cool format, by incorporating the successful termination predicate $\downarrow$. That is, the symbol $\sqrt{}$ is added to the signature as a special constant, and the transition rule $\frac{}{\sqrt{\downarrow}}$ is added to the TSS. Moreover, transition rules that contain an occurrence of a predicate symbol $\overset{a}{\rightarrow} \sqrt{}$ are adapted accordingly.

It is left to the reader to verify that the adapted TSS of ACP_τ with guarded linear recursion is RBB cool. This fact implies that the rooted branching bisimulation equivalence that this TSS induces is a congruence; see Theorem B.4.1. $\square$

Table 5.2 presents axioms for the abstraction operators, modulo rooted branching bisimulation equivalence. The variables x and y in the axioms range over process terms, while v ranges over $A \cup \{\tau\}$. Let $\mathcal{E}_{ACP_\tau}$ denote $\mathcal{E}_{ACP}$ extended with B1,2 and TI1-5.

Table 5.2. Axioms for abstraction operators

TI1	$v \notin I$	$\tau_I(v) = v$
TI2	$v \in I$	$\tau_I(v) = \tau$
TI3		$\tau_I(\delta) = \delta$
TI4		$\tau_I(x + y) = \tau_I(x) + \tau_I(y)$
TI5		$\tau_I(x \cdot y) = \tau_I(x) \cdot \tau_I(y)$

Theorem 5.4.3. $\mathcal{E}_{ACP_\tau} + RDP, RSP$ *is sound for* ACP_τ *with guarded linear recursion, modulo rooted branching bisimulation equivalence.*

Proof. Since rooted branching bisimulation is both an equivalence and a congruence, we only need to check that if $s = t$ is an axiom in $\mathcal{E}_{ACP_\tau} + RDP, RSP$ and σ a closed substitution that maps the variables in s and t to process terms, then $\sigma(s) \underset{rb}{\leftrightarrow} \sigma(t)$. Here, we only provide some intuition for soundness of the axioms in Table 5.2:

- TI1-3 are the defining equations for the abstraction operator τ_I: TI2 says that it renames atomic actions from I into τ, while TI1,3 say that it leaves atomic actions outside I and the deadlock δ unchanged;
- TI4,5 say that in $\tau_I(t)$, all transitions of t labelled with atomic actions from I are renamed into τ.

These intuitions can be made rigorous by means of explicit rooted branching bisimulation relations between the left- and right-hand sides of closed instantiations of TI1-5. $\square$

Exercise 5.4.6. Derive $\tau_{\{b\}}(\langle X \mid X=aY, Y=bX \rangle) = \langle Z \mid Z=aZ \rangle$ from the axiomatisation $\mathcal{E}_{ACP_\tau} + RDP, RSP$.

5.5 An Example with Queues

To give an example of the use of abstraction, we consider two queues of capacity one that are put in sequence: queue Q_1 reads a datum from a channel 1 and sends this datum into channel 3, while queue Q_2 reads a datum from a channel 3 and sends this datum into channel 2. This system can be depicted as follows:

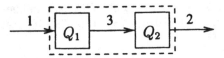

Action $r_i(d)$ represents reading datum d from channel i, while action $s_i(d)$ represents sending datum d into channel i. Q_1 and Q_2 are defined by the recursive specifications

$$Q_1 = \sum_{d \in \Delta} r_1(d) \cdot s_3(d) \cdot Q_1$$
$$Q_2 = \sum_{d \in \Delta} r_3(d) \cdot s_2(d) \cdot Q_2.$$

Here, Δ denotes a finite set of data elements, and as before $\sum_{d \in \Delta} t(d)$ denotes the alternative composition of process terms $t(d)$ for all elements $d \in \Delta$. In the remainder of this section, for notational convenience, the process terms $\langle Q_1 \mid Q_1 = \sum_{d \in \Delta} r_1(d) s_3(d) Q_1 \rangle$ and $\langle Q_2 \mid Q_2 = \sum_{d \in \Delta} r_3(d) s_2(d) Q_2 \rangle$ are abbreviated to Q_1 and Q_2, respectively.

Action $c_3(d)$ denotes communication of datum d through channel 3. Similar as in Example 3.4.1, the communication function γ is defined by $\gamma(s_3(d), r_3(d)) \triangleq c_3(d)$, while all other communications between atomic actions result to δ. The system that consists of queues Q_2 and Q_1 in parallel is described by the process term

$$\tau_{\{c_3(d) \mid d \in \Delta\}}(\partial_{\{s_3(d), r_3(d) \mid d \in \Delta\}}(Q_2 \| Q_1)).$$

The encapsulation operator enforces send and read actions over channel 3 into communication, while the abstraction operator makes internal communication actions over channel 3 invisible.

We show algebraically that the two buffers Q_2 and Q_1 of capacity one in parallel behave as a buffer of capacity two, which can read two data elements from channel 1 before sending them into channel 2. The buffer of capacity two over Δ is described by the linear recursive specification

$$X \qquad = \sum_{d \in \Delta} r_1(d) \cdot Y(d)$$
$$Y(d) \quad = \sum_{d' \in \Delta} r_1(d') \cdot Z(d, d') + s_2(d) \cdot X$$
$$Z(d, d') = s_2(d) \cdot Y(d').$$

In state X, the buffer of capacity two is empty, so that it can only read a datum d from channel 1 and proceed to the state $Y(d)$ where the buffer contains d. In $Y(d)$, the buffer can either read a second datum d' from channel 1 and proceed to the state $Z(d, d')$ where the buffer contains d and d', or send datum d into channel 2 and proceed to the state X where the buffer is empty. Finally, in state $Z(d, d')$ the buffer is full, so that it can only send datum d into channel 2 and proceed to the state $Y(d')$ where it contains d'.

In order to simplify the presentation, we assume that the data set Δ consists of the single element 0, and atomic actions are abbreviated by omitting the suffix (0). We proceed to derive that $\tau_{\{c_3\}}(\partial_{\{s_3,r_3\}}(Q_2 \| Q_1))$ is a solution for X in the recursive specification for the buffer of capacity two. First we expand $\partial_{\{s_3,r_3\}}(Q_2 \| Q_1)$; in each derivation step, the subterms that are reduced are underlined. Since $\gamma(r_3, r_1) \equiv \delta$, the axioms in $\mathcal{E}_{\mathrm{ACP}}$ together with RDP yield:

$$
\begin{aligned}
& \underline{Q_2 \| Q_1} \\
\overset{\mathrm{M1}}{=}\ & \underline{Q_2 \mathbin{\underline{\|}} Q_1} + \underline{Q_1 \mathbin{\underline{\|}} Q_2} + \underline{Q_2 | Q_1} \\
\overset{\mathrm{RDP}}{=}\ & (r_3 s_2 Q_2) \mathbin{\underline{\|}} Q_1 + (r_1 s_3 Q_1) \mathbin{\underline{\|}} Q_2 + (r_3 s_2 Q_2) | (r_1 s_3 Q_1) \\
\overset{\mathrm{LM3,CM8}}{=}\ & r_3((s_2 Q_2) \| Q_1) + r_1((s_3 Q_1) \| Q_2) + \underline{\delta((s_2 Q_2) \| (s_3 Q_1))} \\
\overset{\mathrm{A7}}{=}\ & r_3((s_2 Q_2) \| Q_1) + \underline{r_1((s_3 Q_1) \| Q_2) + \delta} \\
\overset{\mathrm{A6}}{=}\ & r_3((s_2 Q_2) \| Q_1) + r_1((s_3 Q_1) \| Q_2).
\end{aligned}
$$

So the axioms for deadlock and encapsulation yield:

$$
\begin{aligned}
& \partial_{\{s_3,r_3\}}(Q_2 \| Q_1) \\
=\ & \partial_{\{s_3,r_3\}}(r_3((s_2 Q_2) \| Q_1) + r_1((s_3 Q_1) \| Q_2)) \\
\overset{\mathrm{D4}}{=}\ & \underline{\partial_{\{s_3,r_3\}}(r_3((s_2 Q_2) \| Q_1))} + \underline{\partial_{\{s_3,r_3\}}(r_1((s_3 Q_1) \| Q_2))} \\
\overset{\mathrm{D5}}{=}\ & \underline{\partial_{\{s_3,r_3\}}(r_3)}\partial_{\{s_3,r_3\}}((s_2 Q_2) \| Q_1) + \underline{\partial_{\{s_3,r_3\}}(r_1)}\partial_{\{s_3,r_3\}}((s_3 Q_1) \| Q_2) \\
\overset{\mathrm{D1,2}}{=}\ & \underline{\delta \partial_{\{s_3,r_3\}}((s_2 Q_2) \| Q_1)} + r_1 \partial_{\{s_3,r_3\}}((s_3 Q_1) \| Q_2) \\
\overset{\mathrm{A7}}{=}\ & \underline{\delta + r_1 \partial_{\{s_3,r_3\}}((s_3 Q_1) \| Q_2)} \\
\overset{\mathrm{A6}}{=}\ & r_1 \partial_{\{s_3,r_3\}}((s_3 Q_1) \| Q_2).
\end{aligned}
$$

Summarising, we have derived

$$
\partial_{\{s_3,r_3\}}(Q_2 \| Q_1) = r_1 \partial_{\{s_3,r_3\}}((s_3 Q_1) \| Q_2). \tag{5.6}
$$

We proceed to expand $\partial_{\{s_3,r_3\}}((s_3 Q_1) \| Q_2)$. As above, it can be derived from the axioms in $\mathcal{E}_{\mathrm{ACP}}$ together with RDP that

$$
(s_3 Q_1) \| Q_2 = s_3(Q_1 \| Q_2) + r_3((s_2 Q_2) \| (s_3 Q_1)) + c_3(Q_1 \| (s_2 Q_2)).
$$

Using the equation above, it can be derived from the axioms for deadlock and encapsulation that

$$
\partial_{\{s_3,r_3\}}((s_3 Q_1) \| Q_2) = c_3 \partial_{\{s_3,r_3\}}(Q_1 \| (s_2 Q_2)). \tag{5.7}
$$

We proceed to expand $\partial_{\{s_3,r_3\}}(Q_1\|(s_2Q_2))$. By the axioms in $\mathcal{E}_{\text{ACP}}$ together with RDP,

$$Q_1\|(s_2Q_2) = r_1((s_3Q_1)\|(s_2Q_2)) + s_2(Q_2\|Q_1).$$

So by the axioms for encapsulation,

$$\begin{aligned}
&\partial_{\{s_3,r_3\}}(Q_1\|(s_2Q_2)) \\
&= r_1\partial_{\{s_3,r_3\}}((s_3Q_1)\|(s_2Q_2)) + s_2\partial_{\{s_3,r_3\}}(Q_2\|Q_1).
\end{aligned} \tag{5.8}$$

We proceed to expand $\partial_{\{s_3,r_3\}}((s_3Q_1)\|(s_2Q_2))$. By the axioms in $\mathcal{E}_{\text{ACP}}$ together with RDP,

$$(s_3Q_1)\|(s_2Q_2) = s_3(Q_1\|(s_2Q_2)) + s_2(Q_2\|(s_3Q_1)).$$

So by the axioms for deadlock and encapsulation,

$$\partial_{\{s_3,r_3\}}((s_3Q_1)\|(s_2Q_2)) = s_2\partial_{\{s_3,r_3\}}(Q_2\|(s_3Q_1)).$$

Commutativity of the merge with respect to bisimulation equivalence (cf. the second case of Exercise 3.2.1) together with completeness of $\mathcal{E}_{\text{ACP}}+\text{RDP}$, RSP for ACP with linear recursion modulo bisimulation equivalence (see Theorem 4.4.1) yield $Q_2\|(s_3Q_1) = (s_3Q_1)\|Q_2$, so

$$\partial_{\{s_3,r_3\}}((s_3Q_1)\|(s_2Q_2)) = s_2\partial_{\{s_3,r_3\}}((s_3Q_1)\|Q_2). \tag{5.9}$$

Summarising, we have algebraically derived the following relations:

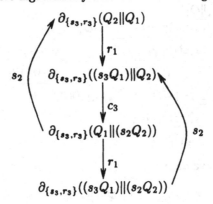

Equations (5.6) and (5.7) together with the axioms for silent step and abstraction yield:

$$\begin{aligned}
\tau_{\{c_3\}}(\partial_{\{s_3,r_3\}}(Q_2\|Q_1)) &\stackrel{(5.6)}{=} \tau_{\{c_3\}}(r_1\partial_{\{s_3,r_3\}}((s_3Q_1)\|Q_2)) \\
&\stackrel{\text{TI1,5}}{=} r_1\tau_{\{c_3\}}(\partial_{\{s_3,r_3\}}((s_3Q_1)\|Q_2)) \\
&\stackrel{(5.7)}{=} r_1\tau_{\{c_3\}}(c_3\partial_{\{s_3,r_3\}}(Q_1\|(s_2Q_2))) \\
&\stackrel{\text{TI2,5}}{=} r_1\tau\tau_{\{c_3\}}(\partial_{\{s_3,r_3\}}(Q_1\|(s_2Q_2))) \\
&\stackrel{\text{B1}}{=} r_1\tau_{\{c_3\}}(\partial_{\{s_3,r_3\}}(Q_1\|(s_2Q_2))).
\end{aligned}$$

Moreover, equation (5.8) together with the axioms for abstraction yield:

$$\tau_{\{c_3\}}(\partial_{\{s_3,r_3\}}(Q_1\|(s_2Q_2)))$$
$$\overset{(5.8)}{=} \tau_{\{c_3\}}(r_1\partial_{\{s_3,r_3\}}((s_3Q_1)\|(s_2Q_2)) + s_2\partial_{\{s_3,r_3\}}(Q_2\|Q_1))$$
$$\overset{\text{TI1,4,5}}{=} r_1\tau_{\{c_3\}}(\partial_{\{s_3,r_3\}}((s_3Q_1)\|(s_2Q_2))) + s_2\tau_{\{c_3\}}(\partial_{\{s_3,r_3\}}(Q_2\|Q_1)).$$

Finally, equations (5.7) and (5.9) together with the axioms for silent step and abstraction yield:

$$\tau_{\{c_3\}}(\partial_{\{s_3,r_3\}}((s_3Q_1)\|(s_2Q_2))) \overset{(5.9)}{=} \tau_{\{c_3\}}(s_2\partial_{\{s_3,r_3\}}((s_3Q_1)\|Q_2))$$
$$\overset{\text{TI1,5}}{=} s_2\tau_{\{c_3\}}(\partial_{\{s_3,r_3\}}((s_3Q_1)\|Q_2))$$
$$\overset{(5.7)}{=} s_2\tau_{\{c_3\}}(c_3\partial_{\{s_3,r_3\}}(Q_1\|(s_2Q_2)))$$
$$\overset{\text{TI2,5}}{=} s_2\tau\tau_{\{c_3\}}(\partial_{\{s_3,r_3\}}(Q_1\|(s_2Q_2)))$$
$$\overset{\text{B1}}{=} s_2\tau_{\{c_3\}}(\partial_{\{s_3,r_3\}}(Q_1\|(s_2Q_2))).$$

The last three derivations together show that

$$X := \tau_{\{c_3\}}(\partial_{\{s_3,r_3\}}(Q_2\|Q_1))$$
$$Y := \tau_{\{c_3\}}(\partial_{\{s_3,r_3\}}(Q_1\|(s_2Q_2)))$$
$$Z := \tau_{\{c_3\}}(\partial_{\{s_3,r_3\}}((s_3Q_1)\|(s_2Q_2)))$$

is a solution for the linear recursive specification E for the buffer of capacity two over $\{0\}$:

$$X = r_1Y$$
$$Y = r_1Z + s_2X$$
$$Z = s_2Y.$$

Hence, by RSP, $\tau_{\{c_3\}}(\partial_{\{s_3,r_3\}}(Q_2\|Q_1)) = \langle X|E\rangle$.

Exercise 5.5.1. Fill in the omitted details of the derivations of equations (5.7), (5.8), and (5.9).

Exercise 5.5.2. Prove that $\tau_{\{c_3\}}(\partial_{\{s_3,r_3\}}(Q_1\|Q_2))$ behaves as a buffer of capacity two. (Hint: this requires one extra application of commutativity of the merge.)

Exercise 5.5.3. Prove that two buffers of capacity one over a finite data set Δ in parallel form a buffer of capacity two over Δ.

5.6 Cluster Fair Abstraction Rule

Though τ-loops are prohibited in guarded linear recursive specifications, they can be constructed using the abstraction operator. For example, the process term $\tau_{\{a\}}(\langle X | X=aX\rangle)$ can only execute τ's until infinity. This observation motivates the following distinction between specifiable and constructible regular processes (see [28]):

- *specifiable* regular processes are the process graphs belonging to process terms in ACP with silent step and guarded linear recursion;
- *constructible* regular processes are the process graphs belonging to process terms in ACP_τ with guarded linear recursion.

$\tau\tau\tau\cdots$ is the simplest example of a regular process that is constructible, being the process graph of $\tau_{\{a\}}(\langle X \mid X=aX\rangle)$, but not specifiable. In general, a constructible regular process is specifiable if and only if it is free of τ-loops. One extra axiom is needed to equate process terms of which the regular process graphs are constructible but not specifiable. For example,

$$\tau_{\{a\}}(\langle X \mid X=aX\rangle) \Leftrightarrow_{rb} \tau_{\{a,b\}}(\langle Y \mid Y=aZ, Z=bY\rangle)$$

because both process terms execute τ's until infinity. However, these process terms cannot be equated by means of $\mathcal{E}_{ACP_\tau} + RDP, RSP$, due to the guardedness restriction on RSP, which is essential for soundness of this axiom. In order to get rid of τ-loops, we introduce the notion of *fair abstraction*. For example, let E denote a guarded linear recursive specification

$$X_1 \quad = aX_2 + s_1$$
$$\vdots$$
$$X_{n-1} = aX_n + s_{n-1}$$
$$X_n \quad = aX_1 + s_n$$

for some $a \in A$. The process term $\tau_{\{a\}}(\langle X_1|E\rangle)$ executes τ-transitions that are the result of abstracting away from the occurrences of a in front of the recursion variables X_i, until it exits this τ-loop by executing one of the process terms $\tau_{\{a\}}(s_i)$ for $i \in \{1,\dots,n\}$. Note that the transitions in the τ-loop are all truly silent, because they do not lose possible behaviours; after the execution of such a τ, it is still possible to execute any of the process terms $\tau_{\{a\}}(s_i)$ for $i \in \{1,\dots,n\}$. Fair abstraction says that $\tau_{\{a\}}(\langle X_1|E\rangle)$ does not stay in the τ-loop forever, so that at some time it will start executing a $\tau_{\{a\}}(s_i)$. Hence,

$$\tau_{\{a\}}(\langle X_1|E\rangle) \Leftrightarrow_{rb} \tau_{\{a\}}(s_1 + \tau(s_1 + \cdots + s_n)).$$

Namely, initially $\tau_{\{a\}}(\langle X_1|E\rangle)$ can execute either $\tau_{\{a\}}(s_1)$ or τ. In the latter case, this initial (so non-silent) τ-transition is followed by the execution of a series of truly silent τ's in the τ-loop, until one of the process terms $\tau_{\{a\}}(s_i)$ for $i \in \{1,\dots,n\}$ is executed.

Exercise 5.6.1. Show that the following pairs of process terms are rooted branching bisimilar:

- $\tau_{\{a\}}(\langle X \mid X=aX\rangle)$ and $\tau\delta$;
- $\tau_{\{a\}}(\langle X \mid X=aX+b\rangle)$ and $b + \tau b$;
- $\tau \cdot \tau_{\{a\}}(\langle X \mid X=aY+b, Y=aX+c\rangle)$ and $\tau(b + c)$.

We proceed to present an axiom to eliminate a cluster of τ-transitions, so that only the exits of such a cluster remains. First, a precise definition is needed of a cluster and its exits.

Definition 5.6.1 (Cluster). *Let E be a guarded linear recursive specification, and $I \subseteq A$. Two recursion variables X and Y in E are in the same cluster for I if and only if there exist sequences of transitions $\langle X|E\rangle \xrightarrow{b_1} \cdots \xrightarrow{b_m} \langle Y|E\rangle$ and $\langle Y|E\rangle \xrightarrow{c_1} \cdots \xrightarrow{c_n} \langle X|E\rangle$ with $b_1, \ldots, b_m, c_1, \ldots, c_n \in I \cup \{\tau\}$.*
a or aX is an exit for the cluster C if and only if:

1. *a or aX is a summand at the right-hand side of the recursive equation for a recursion variable in C; and*
2. *in the case of aX, either $a \notin I \cup \{\tau\}$ or $X \notin C$.*

Exercise 5.6.2. Let E be a guarded linear recursive specification, and $I \subseteq A$. Verify that being in the same cluster for I defines an equivalence relation on the recursion variables in E.

Table 5.3 presents an axiom called *cluster fair abstraction rule* (CFAR) for guarded linear recursive specifications. CFAR allows us to abstract away from a cluster of actions that are renamed into τ, after which only the exits of this cluster remain. CFAR was introduced by Vaandrager [189]; it is a generalisation of a similar principle by Koomen [19, 44, 136]. In Table 5.3, E is a guarded linear recursive specification, $X, Y_1, \ldots, Y_m$ are recursion variables in E, and $v, v_1, \ldots, v_m, w_1, \ldots, w_n$ range over $A \cup \{\tau\}$. Owing to the presence of the initial action v at the left- and right-hand side of CFAR, the initial τ-transitions of $\tau_I(\langle X|E\rangle)$ can be truly silent. If the set of exits is empty, then as always the empty sum at the right-hand side of CFAR represents δ.

Table 5.3. Cluster fair abstraction rule

CFAR If X is in a cluster for I with exits $\{v_1 Y_1, \ldots, v_m Y_m, w_1, \ldots, w_n\}$, then

$$v \cdot \tau_I(\langle X|E\rangle) = v \cdot \tau_I(v_1\langle Y_1|E\rangle + \cdots + v_m\langle Y_m|E\rangle + w_1 + \cdots + w_n)$$

Theorem 5.6.1. *The axiom CFAR is sound modulo rooted branching bisimulation equivalence.*

Proof. Let X be in a cluster for I with exits $\{a_1 Y_1, \ldots, a_m Y_m, b_1, \ldots, b_n\}$. Then $\langle X|E\rangle$ can execute a string of atomic actions from $I \cup \{\tau\}$ inside the cluster of X, followed by an exit $a_i \langle Y_i|E\rangle$ (for some $i \in \{1, \ldots, m\}$) or b_j (for some $j \in \{1, \ldots, n\}$). Hence, $\tau_I(\langle X|E\rangle)$ can execute a string of τ's inside the cluster of X, followed by an exit $\tau_I(a_i\langle Y_i|E\rangle)$ (for some $i \in \{1, \ldots, m\}$) or $\tau_I(b_j)$ (for some $j \in \{1, \ldots, n\}$). The execution of τ's inside the cluster does not lose the possibility to execute any of the exits. Moreover, in the process graph of $a\tau_I(\langle X|E\rangle)$ these τ's are non-initial, owing to the initial a-transition, so they are truly silent. This means that modulo rooted branching bisimulation equivalence only the exits of the cluster of X remain, i.e.,

$$a \cdot \tau_I(\langle X|E\rangle) \; \underline{\leftrightarrow}_{rb} \; a \cdot \tau_I(a_1\langle Y_1|E\rangle + \cdots + a_m\langle Y_m|E\rangle + b_1 + \cdots + b_n).$$

So CFAR is sound modulo rooted branching bisimulation equivalence. $\square$

Example 5.6.1. Let E denote the guarded linear recursive specification

$$X = heads \cdot X + tails.$$

The process term $\langle X|E\rangle$ represents tossing a fair coin until the result is tails. We abstract away from throwing heads, expressed by $\tau_{\{heads\}}(\langle X|E\rangle)$.

$\{X\}$ is the only cluster for $\{heads\}$, and the only exit of this cluster is the atomic action *tails*. So

$$\tau \cdot \tau_{\{heads\}}(\langle X|E\rangle) \overset{\text{CFAR}}{=} \tau \cdot \tau_{\{heads\}}(tails) \overset{\text{TI1}}{=} \tau \cdot tails. \qquad (5.10)$$

Hence,

$$\begin{aligned}
\tau_{\{heads\}}(\langle X|E\rangle) &\overset{\text{RDP}}{=} \tau_{\{heads\}}(heads \cdot \langle X|E\rangle + tails) \\
&\overset{\text{TI1,2,4,5}}{=} \tau \cdot \tau_{\{heads\}}(\langle X|E\rangle) + tails \\
&\overset{(5.10)}{=} \tau \cdot tails + tails.
\end{aligned}$$

In other words, fair abstraction implies that tossing a fair coin infinitely many times will eventually produce the result tails.

Example 5.6.2. We show how to derive the equation

$$\tau_{\{a\}}(\langle X \mid X=aX\rangle) = \tau_{\{a,b\}}(\langle Y \mid Y=aZ, Z=bY\rangle).$$

$\{X\}$ is the only cluster for $\{a\}$ in $\{X=aX\}$, with no exits, so

$$\begin{aligned}
\tau_{\{a\}}(\langle X \mid X=aX\rangle) &\overset{\text{RDP,TI2,5}}{=} \tau \cdot \tau_{\{a\}}(\langle X \mid X=aX\rangle) \\
&\overset{\text{CFAR}}{=} \tau \cdot \tau_{\{a\}}(\delta) \\
&\overset{\text{TI3}}{=} \tau\delta. \qquad (5.11)
\end{aligned}$$

Furthermore, $\{Y, Z\}$ is the only cluster for $\{a, b\}$ in $\{Y=aZ, Z=bY\}$, with no exits, so

$$\begin{aligned}
\tau_{\{a,b\}}(\langle Y \mid Y=aZ, Z=bY\rangle) &\overset{\text{RDP,TI2,5}}{=} \tau \cdot \tau_{\{a,b\}}(\langle Z \mid Y=aZ, Z=bY\rangle) \\
&\overset{\text{CFAR}}{=} \tau \cdot \tau_{\{a,b\}}(\delta) \\
&\overset{\text{TI3}}{=} \tau\delta. \qquad (5.12)
\end{aligned}$$

Hence,

$$\tau_{\{a\}}(\langle X \mid X=aX\rangle) \overset{(5.11)}{=} \tau\delta \overset{(5.12)}{=} \tau_{\{a,b\}}(\langle Y \mid Y=aZ, Z=bY\rangle)$$

The following completeness result is due to van Glabbeek [104].

Theorem 5.6.2. $\mathcal{E}_{\text{ACP}_\tau} + \text{RDP}, \text{RSP}, \text{CFAR}$ *is complete for* ACP_τ *with guarded linear recursion, modulo rooted branching bisimulation equivalence.*

Proof. It suffices to prove that each process term t in ACP_τ with guarded linear recursion is provably equal to a process term $\langle X|E\rangle$ with E a guarded linear recursive specification. Namely, then the desired completeness result follows immediately from the fact that if $\langle X_1|E_1\rangle \underline{\leftrightarrow}_{rb} \langle Y_1|E_2\rangle$ for guarded linear recursive specifications E_1 and E_2, then $\langle X_1|E_1\rangle = \langle Y_1|E_2\rangle$ can be derived from $\mathcal{E}_{\text{ACP}} + \text{B1}, 2 + \text{RDP}, \text{RSP}$; see the proof of Theorem 5.3.2.

We apply structural induction with respect to the size of t. It was shown at the start of the proof of Theorem 5.3.2 that each process term in ACP with silent step and guarded linear recursion is provably equal to a process term $\langle X|E\rangle$ with E a guarded linear recursive specification. So the only case that remains to be covered is when $t \equiv \tau_I(s)$. By induction it may be assumed that $s = \langle X|E\rangle$ with E a guarded linear recursive specification, so $t = \tau_I(\langle X|E\rangle)$. We divide the collection of recursion variables in E into its clusters $C_1, \ldots, C_N$ for I. For $i \in \{1, \ldots, N\}$, let

$$a_{i1}Y_{i1} + \cdots + a_{im_i}Y_{im_i} + b_{i1} + \cdots + b_{in_i}$$

be the alternative composition of exits for the cluster C_i. Furthermore, for atomic actions $a \in A \cup \{\tau\}$ we define

$$\hat{a} = \begin{cases} \tau \text{ if } a \in I \\ a \text{ otherwise.} \end{cases}$$

Finally, for $Z \in C_i$ ($i \in \{1, \ldots, N\}$) we define

$$s_Z \overset{\Delta}{=} \hat{a}_{i1}\tau_I(\langle Y_{i1}|E\rangle) + \cdots + \hat{a}_{im_i}\tau_I(\langle Y_{im_i}|E\rangle) + \hat{b}_{i1} + \cdots + \hat{b}_{in_i}. \quad (5.13)$$

For $Z \in C_i$ and $a \in A \cup \{\tau\}$,

$$a\tau_I(\langle Z|E\rangle) \overset{\text{CFAR}}{=} a\tau_I(a_{i1}\langle Y_{i1}|E\rangle + \cdots + a_{im_i}\langle Y_{im_i}|E\rangle + b_{i1} + \cdots + b_{in_i})$$
$$\overset{\text{T11-5}}{=} as_Z. \quad (5.14)$$

Let the linear recursive specification F contain the same recursion variables as E, where for each $Z \in C_i$ the recursive equation in F is

$$Z = \hat{a}_{i1}Y_{i1} + \cdots + \hat{a}_{im_i}Y_{im_i} + \hat{b}_{i1} + \cdots + \hat{b}_{in_i}.$$

We show that there is no sequence of one or more τ-transitions from $\langle Z|F\rangle$ to itself. Suppose $\hat{a}_{ij} \equiv \tau$ for some $j \in \{1, \ldots, m_i\}$. Then the fact that $a_{ij}Y_{ij}$ is an exit for the cluster C_i ensures that $Y_{ij} \notin C_i$, so there cannot exist a sequence of transitions $\langle Y_{ij}|E\rangle \overset{d_1}{\to} \cdots \overset{d_\ell}{\to} \langle Z|E\rangle$ with $d_1, \ldots, d_\ell \in I \cup \{\tau\}$.

Then by the definition of F there cannot exist a sequence of transitions $\langle Y_{ij}|F\rangle \xrightarrow{\tau} \cdots \xrightarrow{\tau} \langle Z|F\rangle$. Hence, F is guarded.

For each recursion variable $Z \in C_i$ ($i \in \{1, \ldots, N\}$),

$$s_Z \overset{(5.13),(5.14)}{=} \hat{a}_{i1}s_{Y_{i1}} + \cdots + \hat{a}_{im_i}s_{Y_{im_i}} + \hat{b}_{i1} + \cdots + \hat{b}_{in_i}.$$

This means that substituting s_Z for recursion variables Z in F is a solution for F. Hence, by RSP, $s_Z = \langle Z|F\rangle$ for recursion variables Z in F. So for $a \in A \cup \{\tau\}$ and recursion variables Z in F,

$$a\tau_I(\langle Z|E\rangle) \overset{(5.14)}{=} as_Z = a\langle Z|F\rangle. \tag{5.15}$$

Recall that $t = \tau_I(\langle X|E\rangle)$. Let the linear recursive equation for X in E be

$$X = c_1Z_1 + \cdots + c_kZ_k + d_1 + \cdots + d_\ell.$$

Let the linear recursive specification G consist of F extended with a fresh recursion variable W and the recursive equation

$$W = \hat{c}_1Z_1 + \cdots + \hat{c}_kZ_k + \hat{d}_1 + \cdots + \hat{d}_\ell.$$

Since F is guarded, it is clear that G is also guarded.

$$\begin{aligned}
\tau_I(\langle X|E\rangle) &\overset{\text{RDP}}{=} \tau_I(c_1\langle Z_1|E\rangle + \cdots + c_k\langle Z_k|E\rangle + d_1 + \cdots + d_\ell) \\
&\overset{\text{TI1-5}}{=} \hat{c}_1\tau_I(\langle Z_1|E\rangle) + \cdots + \hat{c}_k\tau_I(\langle Z_k|E\rangle) + \hat{d}_1 + \cdots + \hat{d}_\ell \\
&\overset{(5.15)}{=} \hat{c}_1\langle Z_1|F\rangle + \cdots + \hat{c}_k\langle Z_k|F\rangle + \hat{d}_1 + \cdots + \hat{d}_\ell.
\end{aligned}$$

Furthermore, for $Z \in C_i$ ($i \in \{1, \ldots, N\}$),

$$\langle Z|F\rangle \overset{\text{RDP}}{=} \hat{a}_{i1}\langle Y_{i1}|F\rangle + \cdots + \hat{a}_{im_i}\langle Y_{im_i}|F\rangle + \hat{b}_{i1} + \cdots + \hat{b}_{in_i}.$$

Hence, substituting $\tau_I(\langle X|E\rangle)$ for W and $\langle Z|F\rangle$ for all other recursion variables Z in G is a solution for G. So RSP yields

$$\tau_I(\langle X|E\rangle) = \langle W|G\rangle. \qquad \square$$

Exercise 5.6.3. Derive the following equations from the axioms:

- $\tau_{\{a\}}(\langle X \mid X{=}aX{+}b\rangle) = \tau_{\{a\}}(\langle Y \mid Y{=}aZ{+}b, Z{=}aY\rangle)$;
- $\tau_{\{a\}}(\langle X \mid X{=}aY, Y{=}aX{+}bX\rangle) = \langle V \mid V{=}\tau W, W{=}bV\rangle$;
- $\tau_{\{a\}}(\langle X \mid X{=}aY{+}b, Y{=}aX{+}c\rangle) = \tau(b + c) + b$;
- $\tau \cdot \tau_{\{a\}}(\langle X \mid X{=}aY{+}bY, Y{=}aX{+}cX\rangle) = \tau \cdot \langle Z \mid Z{=}bZ{+}cZ\rangle$.

6. Protocol Verifications

Chapters 2-5 presented a standard framework ACP_τ with guarded linear recursion for the specification and manipulation of concurrent processes. Summarising, it consists of basic operators $(A, +, \cdot)$ to define finite processes, communication operators $(\|, \mathbb{L}, |)$ to express parallelism, deadlock and encapsulation (δ, ∂_H) to force atomic actions into communication, silent step and abstraction (τ, τ_I) to make internal computations invisible, and guarded linear recursion $(\langle X|E \rangle)$ to capture regular processes. These constructs form a solid basis for the analysis of a wide range of systems.

In particular, the framework is suitable for the specification and verification of network protocols. For such a verification, the desired external behaviour of the protocol is represented in the form of a process term that is built from the basic operators of BPA together with linear recursion. Moreover, the implementation of the protocol is represented in the form of a process term that involves the basic operators, the three parallel operators, and linear recursion. Next, the internal send and read actions of the implementation are forced into communication using an encapsulation operator, and the internal communication actions are made invisible using an abstraction operator, so that only the input/output relation of the implementation remains. Finally, if the two resulting process terms can be equated by $\mathcal{E}_{ACP_\tau} + \text{RDP}, \text{RSP}, \text{CFAR}$, then this proves that the process graphs belonging to the desired external behaviour and to the input/output relation of the implementation are rooted branching bisimilar.

6.1 Alternating Bit Protocol

Suppose two armies have agreed to attack a city at the same time. The two armies reside on different hills, while the city lies in between these two hills. The only way for the armies to communicate with each other is by sending messengers through the hostile city. This communication is inherently unsafe; if a messenger is caught inside the city, then the message does not reach its destination. The paradox is that in such a situation, the two armies are never able to be 100% sure that they have agreed on a time to attack the city. Namely, if one army sends the message that it will attack at say 11am, then the other army has to acknowledge reception of this message, army one has to acknowledge the reception of this acknowledgement, et cetera.

The alternating bit protocol (ABP) [31], which was already described in the introduction, is a method to ensure successful transmission of data through a corrupted channel (such as messengers through a hostile city). This success is based on the assumption that data can be resent an unlimited number of times. The protocol is depicted in Fig. 6.1.

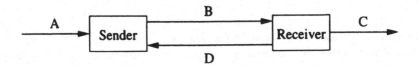

Fig. 6.1. Alternating bit protocol

Data elements $d_1, d_2, d_3, \ldots$ from a finite set Δ are communicated between a Sender and a Receiver. If the Sender reads a datum from channel A, then this datum is communicated through channel B to the Receiver, which sends the datum into channel C. However, channel B is corrupted, so that a message that is communicated through this channel can be turned into an error message $\perp$. Therefore, every time the Receiver receives a message via channel B, it sends an acknowledgement to the Sender via channel D, which is also corrupted.

In the ABP, the Sender attaches a bit 0 to data elements d_{2k-1} and a bit 1 to data elements d_{2k}, when they are sent into channel B. As soon as the Receiver reads a datum, it sends back the attached bit via channel D, to acknowledge reception. If the Receiver receives a corrupted message, then it sends the previous acknowledgement to the Sender once more. The Sender keeps on sending a pair (d_i, b) as long as it receives the acknowledgement $1 - b$ or $\perp$. When the Sender receives the acknowledgement b, it starts sending out the next datum d_{i+1} with attached bit $1 - b$, until it receives the acknowledgement $1 - b$, et cetera. Alternation of the attached bit enables the Receiver to determine whether a received datum is really new, and alternation of the acknowledgement enables the Sender to determine whether it acknowledges reception of a datum or of an error message.

We give a linear recursive specification of the ABP in process algebra. Furthermore, we present an algebraic proof that the resulting process term displays the desired external behaviour; that is, the data elements that are read from channel A by the Sender are sent into channel C by the Receiver in the same order, and no data elements are lost. In other words, the process term is a solution for the recursive specification

$$X = \sum_{d \in \Delta} r_A(d) \cdot s_C(d) \cdot X$$

where action $r_A(d)$ represents "read datum d from channel A", and action $s_C(d)$ represents "send datum d into channel C". The verification of the ABP in this section is based on [44] (see also [28]). In comparison to [44], the modelling of the protocol in this section has been simplified in the sense that there are no explicit atomic actions to represent the non-deterministic behaviour of the communication channels in passing on or corrupting data (see Exercise 6.1.3). An alternative verification can be found in [140, 155].

First, we specify the Sender in the state that it is going to send out a datum with the bit b attached to it, represented by the recursion variable S_b for $b \in \{0,1\}$:

$$S_b = \sum_{d \in \Delta} r_A(d) \cdot T_{db}$$
$$T_{db} = (s_B(d,b) + s_B(\perp)) \cdot U_{db}$$
$$U_{db} = r_D(b) \cdot S_{1-b} + (r_D(1-b) + r_D(\perp)) \cdot T_{db}$$

In state S_b, the Sender reads a datum d from channel A. Then it proceeds to state T_{db}, in which it sends datum d into channel B, with the bit b attached to it. However, the pair (d,b) may be distorted by the channel, so that it becomes the error message $\perp$. Next, the system proceeds to state U_{db}, in which it expects to receive the acknowledgement b through channel D, ensuring that the pair (d,b) has reached the Receiver unscathed. If the correct acknowledgement b is received, then the system proceeds to state S_{1-b}, in which it is going to send out a datum with the bit $1-b$ attached to it. If the acknowledgement is either the wrong bit $1-b$ or the error message $\perp$, then the system proceeds to state T_{db}, to send the pair (d,b) into channel B once more.

Next, we specify the Receiver in the state that it is expecting to receive a datum with the bit b attached to it, represented by the recursion variable R_b for $b \in \{0,1\}$:

$$R_b = \sum_{d' \in \Delta} \{r_B(d',b) \cdot s_C(d') \cdot Q_b + r_B(d',1-b) \cdot Q_{1-b}\} + r_B(\perp) \cdot Q_{1-b}$$
$$Q_b = (s_D(b) + s_D(\perp)) \cdot R_{1-b}$$

In state R_b there are two possibilities.

1. If in R_b the Receiver reads a pair (d',b) from channel B, then this constitutes new information, so the datum d' is sent into channel C. Then the Receiver proceeds to state Q_b, in which it sends acknowledgement b to the Sender via channel D. However, this acknowledgement may be distorted by the channel, so that it becomes the error message $\perp$. Next, the Receiver proceeds to state R_{1-b}, in which it is expecting to receive a datum with the bit $1-b$ attached to it.
2. If in R_b the Receiver reads a pair $(d',1-b)$ or an error message $\perp$ from channel B, then this does not constitute new information. So then the

Receiver proceeds to state Q_{1-b} straight away, to send acknowledgement $1-b$ to the Sender via channel D. However, this acknowledgement may be distorted by the channel, so that it becomes the error message $\perp$. Next, the Receiver proceeds to state R_b again.

A send and a read action of the same message $((d,b),\ b,\ \text{or}\ \perp)$ over the same internal channel (B or D) communicate with each other:

$$\gamma(s_B(d,b), r_B(d,b)) \triangleq c_B(d,b)$$
$$\gamma(s_B(\perp), r_B(\perp)) \triangleq c_B(\perp)$$
$$\gamma(s_D(b), r_D(b)) \triangleq c_D(b)$$
$$\gamma(s_D(\perp), r_D(\perp)) \triangleq c_D(\perp)$$

for $d \in \Delta$ and $b \in \{0,1\}$. All other communications between atomic actions result to δ.

The recursive specification E of the ABP, consisting of the recursive equations for the recursion variables S_b, T_{db}, U_{db}, R_b, and Q_b for $d \in \Delta$ and $b \in \{0,1\}$, can easily be transformed into linear form by introducing extra recursion variables to represent $s_C(d') \cdot Q_b$ for $d' \in \Delta$ and $b \in \{0,1\}$. In the remainder of this section, for notational convenience, process terms $\langle X|E \rangle$ are abbreviated to X. The desired concurrent system is obtained by putting R_0 and S_0 in parallel, encapsulating send and read actions over the internal channels B and D, and abstracting away from communication actions over these channels. That is, the ABP is expressed by the process term

$$\tau_I(\partial_H(R_0 \| S_0))$$

with

$$H = \{s_B(d,b), r_B(d,b), s_D(b), r_D(b) \mid d \in \Delta, b \in \{0,1\}\}$$
$$\cup \{s_B(\perp), r_B(\perp), s_D(\perp), r_D(\perp)\}$$
$$I = \{c_B(d,b), c_D(b) \mid d \in \Delta, b \in \{0,1\}\} \cup \{c_B(\perp), c_D(\perp)\}.$$

Before indulging in the formal proof that the ABP is correct, first we explain the behaviour of the process term $\partial_H(R_0 \| S_0)$ on a more intuitive level; its process graph is depicted in Fig. 6.2. Initially, in state 1, a datum d is read from channel A, resulting in state 2. Then an error message $\perp$ is communicated through channel B zero or more times, each time invoking an incorrect acknowledgement 1 or $\perp$. Finally, the pair $(d,0)$ is communicated through channel B, resulting in state 4. Then datum d is sent into channel C, to reach state 5. The corrupted acknowledgement $\perp$ is communicated through channel D zero or more times, each time invoking a renewed attempt to communicate the pair $(d,0)$ through channel B. Finally, acknowledgement 0 is communicated through channel D, resulting in state 7. There the same process is repeated, with the distinction that the bit 1 attached to the datum that is communicated through channel B. Note that states 2-6 and 8-12 depend on the datum d that is read from channel A.

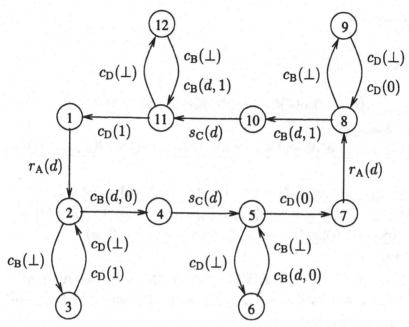

Fig. 6.2. Transition graph of $\partial_H(R_0\|S_0)$.

We proceed with the formal verification of the ABP. First, we derive from $\mathcal{E}_{\text{ACP}_\tau}$ and RDP the six equations I-VI below, which establish the transitions between states 1-7 in the bottom half of Fig. 6.2.

$$\text{I}: \partial_H(R_0\|S_0) \qquad = \sum_{d\in\Delta} r_A(d) \cdot \partial_H(T_{d0}\|R_0)$$

$$\text{II}: \partial_H(T_{d0}\|R_0) \qquad = c_B(d,0) \cdot \partial_H(U_{d0}\|(s_C(d)Q_0)) \\ \qquad\qquad + c_B(\perp) \cdot \partial_H(U_{d0}\|Q_1)$$

$$\text{III}: \partial_H(U_{d0}\|Q_1) \qquad = (c_D(1) + c_D(\perp)) \cdot \partial_H(T_{d0}\|R_0)$$

$$\text{IV}: \partial_H(U_{d0}\|(s_C(d)Q_0)) = s_C(d) \cdot \partial_H(Q_0\|U_{d0})$$

$$\text{V}: \partial_H(Q_0\|U_{d0}) \qquad = c_D(0) \cdot \partial_H(R_1\|S_1) + c_D(\perp) \cdot \partial_H(R_1\|T_{d0})$$

$$\text{VI}: \partial_H(R_1\|T_{d0}) \qquad = (c_B(d,0) + c_B(\perp)) \cdot \partial_H(Q_0\|U_{d0})$$

We start with the derivation of equation I. The process term $R_0\|S_0$ can be expanded as follows. In each step, the subterms that are reduced are underlined.

$$\underline{R_0 \| S_0}$$

$$\overset{\text{M1}}{=}$$

$$\underline{R_0 \mathbin{\underline{\|}} S_0} + \underline{S_0 \mathbin{\underline{\|}} R_0} + \underline{R_0 | S_0}$$

$$\overset{\text{RDP}}{=}$$

$$\underline{(\textstyle\sum_{d' \in \Delta}\{r_B(d',0)s_C(d')Q_0 + r_B(d',1)Q_1\} + r_B(\bot)Q_1) \mathbin{\underline{\|}} S_0}$$
$$+\underline{(\textstyle\sum_{d \in \Delta} r_A(d)T_{d0}) \mathbin{\underline{\|}} R_0}$$
$$+\underline{(\textstyle\sum_{d' \in \Delta}\{r_B(d',0)s_C(d')Q_0 + r_B(d',1)Q_1\} + r_B(\bot)Q_1)|(\textstyle\sum_{d \in \Delta} r_A(d)T_{d0})}$$

$$\overset{\text{LM4,CM9,10}}{=}$$

$$\textstyle\sum_{d' \in \Delta}\{\underline{(r_B(d',0)s_C(d')Q_0) \mathbin{\underline{\|}} S_0} + \underline{(r_B(d',1)Q_1) \mathbin{\underline{\|}} S_0}\} + \underline{(r_B(\bot)Q_1) \mathbin{\underline{\|}} S_0}$$
$$+ \textstyle\sum_{d \in \Delta} \underline{(r_A(d)T_{d0}) \mathbin{\underline{\|}} R_0} + \textstyle\sum_{d' \in \Delta} \textstyle\sum_{d \in \Delta}\{\underline{(r_B(d',0)s_C(d')Q_0)|(r_A(d)T_{d0})}$$
$$+ \underline{(r_B(d',1)Q_1)|(r_A(d)T_{d0})}\} + \textstyle\sum_{d \in \Delta} \underline{(r_B(\bot)Q_1)|(r_A(d)T_{d0})}$$

$$\overset{\text{LM3,CM8}}{=}$$

$$\textstyle\sum_{d' \in \Delta}\{r_B(d',0)((s_C(d')Q_0)\|S_0) + r_B(d',1)(Q_1\|S_0)\} + r_B(\bot)(Q_1\|S_0)$$
$$+ \textstyle\sum_{d \in \Delta} r_A(d)(T_{d0}\|R_0) + \textstyle\sum_{d' \in \Delta} \textstyle\sum_{d \in \Delta}\{\underline{\delta((s_C(d')Q_0)\|T_{d0})} + \underline{\delta(Q_1\|T_{d0})}\}$$
$$+ \textstyle\sum_{d \in \Delta} \underline{\delta(Q_1\|T_{d0})}$$

$$\overset{\text{A6,7}}{=}$$

$$\textstyle\sum_{d' \in \Delta}\{r_B(d',0)((s_C(d')Q_0)\|S_0) + r_B(d',1)(Q_1\|S_0)\} + r_B(\bot)(Q_1\|S_0)$$
$$+ \textstyle\sum_{d \in \Delta} r_A(d)(T_{d0}\|R_0).$$

Next, we expand the process term $\partial_H(R_0\|S_0)$.

$$\partial_H(R_0\|S_0)$$

$$=$$

$$\partial_H(\textstyle\sum_{d' \in \Delta}\{r_B(d',0)((s_C(d')Q_0)\|S_0) + r_B(d',1)(Q_1\|S_0)\}$$
$$+ r_B(\bot)(Q_1\|S_0) + \textstyle\sum_{d \in \Delta} r_A(d)(T_{d0}\|R_0))$$

$$\overset{\text{D4}}{=}$$

$$\textstyle\sum_{d' \in \Delta}\{\partial_H(r_B(d',0)((s_C(d')Q_0)\|S_0)) + \partial_H(r_B(d',1)(Q_1\|S_0))\}$$
$$+ \partial_H(r_B(\bot)(Q_1\|S_0)) + \textstyle\sum_{d \in \Delta} \partial_H(r_A(d)(T_{d0}\|R_0))$$

$$\overset{\text{D1,2,5}}{=}$$

$$\textstyle\sum_{d' \in \Delta}\{\underline{\delta\partial_H((s_C(d')Q_0)\|S_0)} + \underline{\delta\partial_H(Q_1\|S_0)}\} + \underline{\delta\partial_H(Q_1\|S_0)}$$
$$+ \textstyle\sum_{d \in \Delta} r_A(d)\partial_H(T_{d0}\|R_0)$$

$$\overset{\text{A6,7}}{=}$$

$$\textstyle\sum_{d \in \Delta} r_A(d)\partial_H(T_{d0}\|R_0).$$

This completes the proof of equation I. Similar to equation I, we can derive the remaining equations II-VI. These derivations are sketched below.

$$T_{d0}\|R_0 = (s_B(d,0) + s_B(\bot))(U_{d0}\|R_0)$$
$$+ \sum_{d'\in\Delta}\{r_B(d',0)((s_C(d')Q_0)\|T_{d0}) + r_B(d',1)(Q_1\|T_{d0})\}$$
$$+ r_B(\bot)(Q_1\|T_{d0})$$
$$+ c_B(d,0)(U_{d0}\|(s_C(d)Q_0))$$
$$+ c_B(\bot)(U_{d0}\|Q_1)$$

$$\partial_H(T_{d0}\|R_0) = c_B(d,0)\partial_H(U_{d0}\|(s_C(d)Q_0)) + c_B(\bot)\partial_H(U_{d0}\|Q_1)$$

$$U_{d0}\|Q_1 = r_D(0)(S_1\|Q_1)$$
$$+ (r_D(1) + r_D(\bot))(T_{d0}\|Q_1)$$
$$+ (s_D(1) + s_D(\bot))(R_0\|U_{d0})$$
$$+ (c_D(1) + c_D(\bot))(T_{d0}\|R_0)$$

$$\partial_H(U_{d0}\|Q_1) = (c_D(1) + c_D(\bot))\partial_H(T_{d0}\|R_0)$$

$$U_{d0}\|(s_C(d)Q_0) = r_D(0)(S_1\|(s_C(d)Q_0))$$
$$+ (r_D(1) + r_D(\bot))(T_{d0}\|(s_C(d)Q_0))$$
$$+ s_C(d)(Q_0\|U_{d0})$$

$$\partial_H(U_{d0}\|(s_C(d)Q_0)) = s_C(d)\partial_H(Q_0\|U_{d0})$$

$$Q_0\|U_{d0} = (s_D(0) + s_D(\bot))(R_1\|U_{d0})$$
$$+ r_D(0)(S_1\|Q_0)$$
$$+ (r_D(1) + r_D(\bot))(T_{d0}\|Q_0)$$
$$+ c_D(0)(R_1\|S_1)$$
$$+ c_D(\bot)(R_1\|T_{d0})$$

$$\partial_H(Q_0\|U_{d0}) = c_D(0)\partial_H(R_1\|S_1) + c_D(\bot)\partial_H(R_1\|T_{d0})$$

$$R_1\|T_{d0} = \sum_{d'\in\Delta}\{r_B(d',1)((s_C(d')Q_1)\|T_{d0}) + r_B(d',0)(Q_0\|T_{d0})\}$$
$$+ r_B(\bot)(Q_0\|T_{d0})$$
$$+ (s_B(d,0) + s_B(\bot))(U_{d0}\|R_1)$$
$$+ (c_B(d,0) + c_B(\bot))(Q_0\|U_{d0})$$

$$\partial_H(R_1\|T_{d0}) = (c_B(d,0) + c_B(\bot))\partial_H(Q_0\|U_{d0})$$

Note that the process term $\partial_H(R_1\|S_1)$ in the right-hand side of equation V is not the left-hand side of an equation I-VI. We proceed to expand $\partial_H(R_1\|S_1)$. That is, similar to equations I-VI, the following six equations VII-XII can be derived, which establish the transitions between states 7-12 and 1 in the top half of Fig. 6.2. The derivations of these equations are left to the reader.

$$\text{VII}: \partial_H(R_1\|S_1) = \sum_{d\in\Delta} r_A(d)\cdot\partial_H(T_{d1}\|R_1)$$

$$\text{VIII}: \partial_H(T_{d1}\|R_1) = c_B(d,1)\cdot\partial_H(U_{d1}\|(s_C(d)Q_1)) + c_B(\perp)\cdot\partial_H(U_{d1}\|Q_0)$$

$$\text{IX}: \partial_H(U_{d1}\|Q_0) = (c_D(0)+c_D(\perp))\cdot\partial_H(T_{d1}\|R_1)$$

$$\text{X}: \partial_H(U_{d1}\|(s_C(d)Q_1)) = s_C(d)\cdot\partial_H(Q_1\|U_{d1})$$

$$\text{XI}: \partial_H(Q_1\|U_{d1}) = c_D(1)\cdot\partial_H(R_0\|S_0)+c_D(\perp)\cdot\partial_H(R_0\|T_{d1})$$

$$\text{XII}: \partial_H(R_0\|T_{d1}) = (c_B(d,1)+c_B(\perp))\cdot\partial_H(Q_1\|U_{d1})$$

Thus, we have derived algebraically the relations depicted in Fig. 6.2. Owing to equations I-XII, RSP yields

$$\partial_H(R_0\|S_0) = \langle X_1|E\rangle \tag{6.1}$$

where E denotes the linear recursive specification

$$\{ \begin{aligned}
X_1 &= \sum_{d'\in\Delta} r_A(d')\cdot X_{2d'}, & Y_1 &= \sum_{d'\in\Delta} r_A(d')\cdot Y_{2d'},\\
X_{2d} &= c_B(d,0)\cdot X_{4d}+c_B(\perp)\cdot X_{3d}, & Y_{2d} &= c_B(d,1)\cdot Y_{4d}+c_B(\perp)\cdot Y_{3d},\\
X_{3d} &= (c_D(1)+c_D(\perp))\cdot X_{2d}, & Y_{3d} &= (c_D(0)+c_D(\perp))\cdot Y_{2d},\\
X_{4d} &= s_C(d)\cdot X_{5d}, & Y_{4d} &= s_C(d)\cdot Y_{5d},\\
X_{5d} &= c_D(0)\cdot Y_1+c_D(\perp)\cdot X_{6d}, & Y_{5d} &= c_D(1)\cdot X_1+c_D(\perp)\cdot Y_{6d},\\
X_{6d} &= (c_B(d,0)+c_B(\perp))\cdot X_{5d}, & Y_{6d} &= (c_B(d,1)+c_B(\perp))\cdot Y_{5d}
\end{aligned}$$
$$| \ d\in\Delta \ \}.$$

We proceed to prove that the process term $\tau_I(\langle X_1|E\rangle)$ exhibits the desired external behaviour of the ABP. After application of the abstraction operator τ_I to the process term $\langle X_1|E\rangle$, the loops of communication actions in Fig. 6.2 (between states 2-3, states 5-6, states 8-9, and states 11-12) become τ-loops. These loops can be removed using CFAR. For example, for $d\in\Delta$ the recursion variables X_{2d} and X_{3d} form a cluster for I with exit $c_B(d,0)\cdot X_{4d}$, so

$$r_A(d)\cdot\tau_I(\langle X_{2d}|E\rangle) \overset{\text{CFAR}}{=} r_A(d)\cdot\tau_I(c_B(d,0)\langle X_{4d}|E\rangle)$$
$$\overset{\text{TI2,5,B1}}{=} r_A(d)\cdot\tau_I(\langle X_{4d}|E\rangle). \tag{6.2}$$

Similarly, CFAR together with TI2,5 and B1 can be applied to eliminate the other three loops of communication actions. Thus, we derive the following equations:

$$s_C(d)\cdot\tau_I(\langle X_{5d}|E\rangle) = s_C(d)\cdot\tau_I(\langle Y_1|E\rangle) \tag{6.3}$$
$$r_A(d)\cdot\tau_I(\langle Y_{2d}|E\rangle) = r_A(d)\cdot\tau_I(\langle Y_{4d}|E\rangle) \tag{6.4}$$
$$s_C(d)\cdot\tau_I(\langle Y_{5d}|E\rangle) = s_C(d)\cdot\tau_I(\langle X_1|E\rangle). \tag{6.5}$$

Applying RDP, TI1,4,5, and equations (6.2) and (6.3) we derive

$$\tau_I(\langle X_1|E\rangle) \overset{\text{RDP,TI1,4,5}}{=} \sum_{d\in\Delta} r_A(d)\cdot\tau_I(\langle X_{2d}|E\rangle)$$

$$\overset{(6.2)}{=} \sum_{d\in\Delta} r_A(d)\cdot\tau_I(\langle X_{4d}|E\rangle)$$

$$\overset{\text{RDP,TI1,5}}{=} \sum_{d\in\Delta} r_A(d)\cdot s_C(d)\cdot\tau_I(\langle X_{5d}|E\rangle)$$

$$\overset{(6.3)}{=} \sum_{d\in\Delta} r_A(d)\cdot s_C(d)\cdot\tau_I(\langle Y_1|E\rangle). \tag{6.6}$$

Likewise, applying RDP, TI1,4,5, and equations (6.4) and (6.5) we can derive

$$\tau_I(\langle Y_1|E\rangle) = \sum_{d\in\Delta} r_A(d)\cdot s_C(d)\cdot\tau_I(\langle X_1|E\rangle). \tag{6.7}$$

Equations (6.6) and (6.7) together with RSP enable us to derive the following equation (cf. Exercise 4.3.5):

$$\tau_I(\langle X_1|E\rangle) = \sum_{d\in\Delta} r_A(d)\cdot s_C(d)\cdot\tau_I(\langle X_1|E\rangle).$$

In combination with equation (6.1) this yields

$$\tau_I(\partial_H(R_0\|S_0)) = \sum_{d\in\Delta} r_A(d)\cdot s_C(d)\cdot\tau_I(\partial_H(R_0\|S_0)).$$

In other words, the ABP exhibits the desired external behaviour. This finishes the verification of the ABP.

Intuitively, the application of CFAR in the verification excludes the possibility that the channels B and D are completely defective, because a message can only be distorted a finite number of times.

Exercise 6.1.1. Complete the omitted details of the verification of the ABP.

Exercise 6.1.2. Suppose the recursive specification of the Sender in the ABP were adapted as follows:

$$S_b = \sum_{d\in\Delta} r_A(d)\cdot T_{db}$$
$$T_{db} = (s_B(d,b) + s_B(\bot))\cdot U_{db}$$
$$U_{db} = (r_D(b) + r_D(\bot))\cdot S_{1-b} + r_D(1-b)\cdot T_{db}$$

That is, if the Sender receives an acknowledgement $\bot$, then it starts sending the next datum. Show that in that case $\tau_I(\partial_H(R_0\|S_0))$ would not display the desired external behaviour.

Exercise 6.1.3. Let us specify the non-deterministic behaviour of channels B and D. That is, the Sender sends (uncorrupted) data with attached bits into channel B1 and reads messages from channel D2, while the Receiver reads messages from channel B2 and sends (uncorrupted) acknowledgements into channel D1. The processes K and L, which express that messages may be corrupted by channels B and D, respectively, are defined by the recursive equations

$$K = \sum_{d \in \Delta} \sum_{b \in \{0,1\}} r_{B1}(d,b) \cdot (i \cdot s_{B2}(d,b) + i \cdot s_{B2}(\bot)) \cdot K$$

$$L = \sum_{b \in \{0,1\}} r_{D1}(b) \cdot (i \cdot s_{D2}(b) + i \cdot s_{D2}(\bot)) \cdot L$$

The atomic action i does not communicate with any atomic action and is added to the set I. Prove that $\tau_I(\partial_H(R_0 \| S_0 \| K \| L))$ displays the desired external behaviour.

6.2 Bounded Retransmission Protocol

Philips formulated a bounded retransmission protocol BRP for the implementation of a remote control (RC). Data elements that are sent from the RC to their destination, say a TV, may get lost. For example, the user may point the RC in the wrong direction. Therefore, if the TV receives a datum, it sends back a message to the RC, to acknowledge reception; this acknowledgement may also get lost. The RC attaches an alternating bit to each datum that it sends to the TV, so that the TV can recognise whether it received a datum before.

Clearly, there is a strong similarity between the ABP and the BRP. However, there are some fundamental distinctions between the two protocols, which are listed below.

1. In general, the data packets that are sent from the RC to the TV are large, so that they cannot be sent in one go. This means that each data packet is chopped into little pieces, and the RC sends these pieces one by one.
   The RC attaches a special label to the last element of a data packet, so that at reception of this datum the TV recognises that this completes the data packet.
2. In the ABP we took the view that a datum can be resent an unlimited number of times. Owing to this assumption, fair abstraction could be applied to conclude that each datum that is sent by the Sender will eventually reach the Receiver. However, this assumption is not very practical, because here it would mean that the RC could get into an infinite loop, while trying without success to communicate a datum to the TV.

Therefore, Philips requires that a datum can only be resent a limited number of times. This means that the correctness criterion cannot be that each datum that is sent by the RC will eventually reach the TV. Instead, it is required that either the complete data packet is communicated between the RC and the TV, or the RC and the TV send appropriate messages to the outside world to inform their corresponding partners that this communication has (or may have) failed.

3. In the ABP, data does not get lost, but can only be corrupted. This assumption ensures that the protocol always progresses: if the Sender sends a datum to the Receiver, then the Sender will eventually receive either an acknowledgement or an error message. The Sender responds to such a message, which secures that the protocol progresses.

   However, in the communication between the RC and the TV, data elements may get lost. In order to ensure that the BRP progresses, we need to incorporate some notion of time. Namely, if the RC sends a datum to the TV and does not receive an acknowledgement within a certain period of time, then it is certain that the datum or its acknowledgement was lost, so that the datum has to be resent. Furthermore, if the TV does not receive a next datum within a certain period of time, then it can be sure that the RC has given up transmission of a data packet.

   There are a number of ways to add the factor time to process algebra (see Section 6.3 for an explicit method based on timed actions). Here we use two timer processes T_1 and T_2 that send time-out messages to the RC and the TV, respectively. If the RC sends a datum to the TV, then it implicitly sets the timer T_1; if the RC receives an acknowledgement, then it implicitly resets T_1. Alternatively, T_1 sends a time-out to the RC, to signal that the acknowledgement has been delayed for too long; in that case, the RC resends the datum. Likewise, the timer T_2 can send a time-out to the TV, to signal that the next datum has been delayed so long that the RC must have given up transmission of the data packet.

4. In the ABP, an acknowledgement from the Receiver could have been prompted by an error message. Therefore, the Sender required two types of acknowledgements (0 and 1), to distinguish acknowledgements for successful transfers from acknowledgements for error messages.

   In the BRP, data is never corrupted. Hence, when the RC receives an acknowledgement, it can be sure that the TV received the datum unscathed. Therefore, only one kind of acknowledgement is needed.

The BRP is depicted in Fig. 6.3. Note that the medium between the RC and the TV is represented by two separate entities K and L, which can pass on a datum or lose it at random. The dotted lines between these entities and the timer T_1 designate that losing a datum or an acknowledgement triggers T_1 to send a time-out to the RC via channel G. Similarly, the dotted line between the RC and the timer T_2 designates that if the RC gives up transmitting a

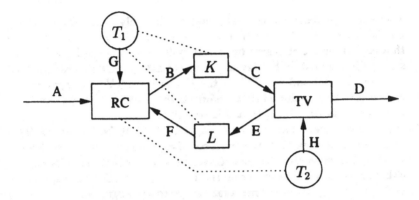

Fig. 6.3. Bounded retransmission protocol

data packet, then this is followed by a delay that is sufficiently long for T_2 to send a time-out to the TV via channel H.

Groote and van de Pol [110] specified the BRP in process algebra, and verified that the protocol exhibits the required external behaviour. First, we give an informal description of the process algebra specification for the BRP, and explain its required external behaviour. Next, we present the formal specification, and derive algebraically its actual external behaviour. Our specification is a simplification of the specification in [110], where setting and resetting the timers is performed by explicit actions, error messages are more sophisticated, and special actions are needed in order to enforce synchronisation of the RC and the TV.

Suppose the RC receives a data packet $(d_1, \ldots, d_N)$ via channel A. Then the RC transmits the data elements $d_1, \ldots, d_N$ separately, where the last datum d_N is supplied with a special label *last*. Furthermore, each datum is supplied with an alternating bit 0 or 1: data elements d_{2k-1} are supplied with bit 0 while data elements d_{2k} are supplied with bit 1. If the RC sends a pair (d_i, b) into channel B for the first time, then it implicitly sets the timer T_1, and moreover it sets a counter at zero to keep track of the number of failed attempts to send datum d_i. Now there are two possibilities:

1. The RC receives an acknowledgement *ack* via channel F. Then it sends out the next pair $(d_{i+1}, 1 - b)$, sets the timer T_1, and gives the counter the value zero.
2. The RC receives a time-out from the timer T_1 via channel G. Then it sends out the pair (d_i, b) again, sets the timer T_1, and increases the value of the counter by one.

Transmission of the data packet is either completed successfully, if the RC receives an acknowledgement from the TV that it received the last datum d_N of the packet, or broken off unsuccessfully, if at some point the counter reaches its preset maximum value *max*. In the first case, the RC sends the message I_{OK} into channel A, to inform the outside world that transmission of the data packet $(d_1, \ldots, d_N)$ was concluded successfully. In the second case, the RC sends the message I_{NOK} into channel A, to inform the outside world that transmission of the data packet failed.

If the TV receives a pair (d_i, b) via channel C for the first time (which can be judged from the bit b), then it sends datum d_i into channel D and acknowledgement ack into channel E. Now there are three possibilities:

1. The TV receives the next pair $(d_{i+1}, 1 - b)$ via channel C. Then it sends d_{i+1} into channel D and ack into channel E.
2. The TV receives the pair (d_i, b) again. Then it only sends ack into channel E.
3. The TV receives a time-out from the timer T_2 via channel H. Then it sends the message I_{NOK} into channel D, to inform its corresponding partner in the outside world that it should ignore the previous messages, because transmission of the data packet was interrupted.

This procedure is repeated until the TV may receive a message $(d_N, b, last)$, in which case it sends both the datum d_N and the message I_{OK} into channel D, to inform its corresponding partner in the outside world that this successfully concludes transmission of the data packet.

K and L represent the non-deterministic behaviour of the medium between the RC and the TV. If K reads a message via channel B, then it may or may not pass on this message to the TV via channel C. In the latter case, the timer T_1 will eventually send a time-out to the RC. Similarly, if L reads a message via channel E, then it may or may not pass on this message to the RC via channel F. In the latter case, the timer T_1 will eventually send a time-out to the RC.

This almost finishes the informal description of the BRP. However, there is one aspect of this protocol that has not yet been discussed, concerning error messages. This characteristic is explained using the specification of the required external behaviour, which is depicted in Fig. 6.4. The clockwise circle in this picture represents successful transfers of data elements (starting at the leftmost node), while the transitions that digress from this circle are error messages.

There are two special cases with respect to error messages that are sent into channels, at the start and at the end of transmission of a data packet.

1. Suppose the first pair $(d_1, 0)$ never reaches the TV, even after the maximum number of tries. Then the RC sends a message I_{NOK} into channel A. However, there is no need for the TV to send such an error message into channel D. Namely, the TV did not send any element of the data packet into channel D, so it does not have to warn its corresponding partner in the outside world.
2. The second special case is quite intricate. Suppose the RC attempted to send the final triple $(d_N, b, last)$ to the TV, but that it did not receive an

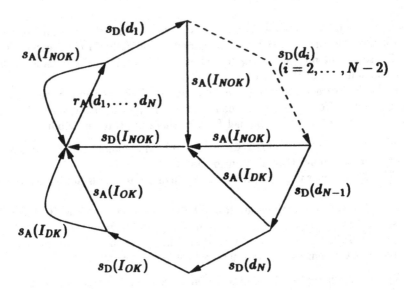

Fig. 6.4. External behaviour of the BRP

acknowledgement, even after the maximum number of tries. Then there are two possibilities:

- Either the TV did receive d_N. In that case the TV received the complete data packet, and it sent the concluding message I_{OK} into channel D.
- Or the TV did not receive d_N. In that case the TV did not receive the complete data packet, and after a time-out from the timer T_2 it will send error message I_{NOK} into channel D.

The RC does not know which of these two cases holds. So it sends a special error message I_{DK} into channel A, to express that it does not know whether transmission of the data packet has been successful.

We proceed to present the recursive equations that formally specify the BRP in process algebra. In order to simplify the specification, we assume that the data packets that reach the RC via channel A have length ≥ 2, and that $max \geq 2$. The recursive specification uses the following data parameters and functions.

- d ranges over a finite data set Δ, and ℓ ranges over the set Λ of lists of data of length ≥ 2. $head(\ell)$ represents the first element of the list ℓ, and $tail(\ell)$ represents the remaining list: $head(d_1, \ldots, d_N) \stackrel{\Delta}{=} d_1$ and $tail(d_1, \ldots, d_N) \stackrel{\Delta}{=} (d_2, \ldots, d_N)$.
- b ranges over $\{0, 1\}$, while n ranges over $\{0, \ldots, max\}$, where max is the maximum number of attempts that the RC is allowed to undertake to transmit a datum to the TV.

- Finally, we have the acknowledgement *ack*, the time-out *to*, the appendix *last* for the last datum of a data packet, and the messages I_{OK}, I_{NOK}, and I_{DK} for the outside world.

We start with the specification of the RC; its initial state is represented by the recursion variable X:

$$
\begin{aligned}
X &= \sum_{\ell \in \Lambda} r_A(\ell) \cdot Y(\ell, 0, 0) \\
Y(\ell, b, n) &= s_B(head(\ell), b) \cdot Z(\ell, b, n) \\
Y(d, b, n) &= s_B(d, b, last) \cdot Z(d, b, n) \\
(n < max) \; Z(\ell, b, n) &= r_F(ack) \cdot Y(tail(\ell), 1 - b, 0) \\
&\quad + r_G(to) \cdot Y(\ell, b, n + 1) \\
Z(\ell, b, max) &= r_F(ack) \cdot Y(tail(\ell), 1 - b, 0) \\
&\quad + r_G(to) \cdot s_A(I_{NOK}) \cdot s_H(to) \cdot X \\
(n < max) \; Z(d, b, n) &= r_F(ack) \cdot s_A(I_{OK}) \cdot X \\
&\quad + r_G(to) \cdot Y(d, b, n + 1) \\
Z(d, b, max) &= r_F(ack) \cdot s_A(I_{OK}) \cdot X \\
&\quad + r_G(to) \cdot s_A(I_{DK}) \cdot s_H(to) \cdot X
\end{aligned}
$$

The intuition behind these recursive equations is as follows. Let l range over lists of data of length ≥ 1.

- In state X, the RC waits until it receives a data packet ℓ via channel A, after which it proceeds to $Y(\ell, 0, 0)$. The first zero represents the bit that is going to be attached to $head(\ell)$, while the second zero represents the counter.
- In state $Y(l, b, n)$, the RC attempts to send the head of list l to the TV via channel B, with bit b attached to it. If l consists of a single datum, then moreover a label *last* is attached to this message. The counter n registers the number of unsuccessful attempts to send the head of l to the TV.
- In state $Z(l, b, n)$, the RC waits for either an acknowledgement via channel F or a time-out via channel G.
  - Suppose the RC receives an acknowledgement from the TV. If l consists of two or more data elements, then it proceeds to send the head of $tail(l)$ to the TV, with bit $1 - b$ attached to it and the counter starting at zero. If l consists of a single datum, then it concludes successful transmission of the data packet by sending I_{OK} into channel A, and proceeds to state X.
  - Suppose the RC receives a time-out from the timer T_1. If $n < max$, then it sends the pair $(head(l), b)$ to the TV again, with the counter increased by one. If $n \equiv max$, then it concludes that transmission of the data packet was unsuccessful (if l consists of two or more elements) or may have been unsuccessful (if l consists of a single element), by sending I_{NOK} or I_{DK} into channel A, respectively. This message is followed by a delay, sufficiently long to let the timer T_2 send a time-out to the TV via channel H, after which the RC proceeds to state X.

Next, we specify the TV; its root state is represented by the recursion variable V:

$$V \quad = \sum_{d \in \Delta} r_C(d, 0) \cdot s_D(d) \cdot s_E(ack) \cdot W(1)$$
$$+ \sum_{d \in \Delta}(r_C(d, 0, last) + r_C(d, 1, last)) \cdot s_E(ack) \cdot V$$
$$+ r_H(to) \cdot V$$

$$W(b) = r_C(d, b) \cdot s_D(d) \cdot s_E(ack) \cdot W(1 - b)$$
$$+ r_C(d, b, last) \cdot s_D(d) \cdot s_D(I_{OK}) \cdot s_E(ack) \cdot V$$
$$+ r_C(d, 1 - b) \cdot s_E(ack) \cdot W(b)$$
$$+ r_H(to) \cdot s_D(I_{NOK}) \cdot V$$

The intuition behind these recursive equations is as follows.

- In state V, the TV is waiting for the first element of a new data packet, with the bit 0 attached to it. If it receives such a message, then it sends the datum into channel D, sends an acknowledgement into channel E, and proceeds to state $W(1)$.

  If the TV receives a message with *last* attached to it, then it recognises that it already received this datum before: it is the last datum of the data packet that it received previously. Hence, the TV only sends an acknowledgement into channel E, and remains in state V.

  Finally, the TV may receive a time-out from the timer T_2 via channel H, which signals that the RC never received an acknowledgement for the last datum of the previous data packet, or that the RC failed to transfer a single datum of some new data packet. Then the TV remains in state V.

- In state $W(b)$, the TV has received some but not all data of a packet from the RC, and is waiting for a datum with the bit b attached to it. If it receives such a message, then it sends the datum into channel D, sends an acknowledgement into channel E, and proceeds to state $W(1 - b)$ to wait for a message with the bit $1 - b$ attached to it. If the TV receives a message with not only b but also *last* attached to it, then it concludes that the data packet has been transferred successfully. In that case it sends both the datum d and the message I_{OK} into channel D, sends an acknowledgement into channel E, and proceeds to state V.

  If the TV receives a message with the bit $1 - b$ attached to it, then it already received this datum before. Hence, it only sends an acknowledgement into channel E, and remains in state $W(b)$.

  Finally, the TV may receive a time-out from the timer T_2 via channel H, which signals that the RC has given up transmission of the data packet. Then the TV sends the error message I_{NOK} into channel D and proceeds to state V.

Finally, we specify the mediums K and L:

$$K = \sum_{d \in \Delta} \sum_{b \in \{0,1\}} \{r_B(d, b) \cdot (s_C(d, b) + s_G(to)) \cdot K$$
$$+ r_B(d, b, last) \cdot (s_C(d, b, last) + s_G(to)) \cdot K\}$$
$$L = r_E(ack) \cdot (s_F(ack) + s_G(to)) \cdot L$$

The intuition behind these recursive equations is as follows.

- If K receives a message from the RC via channel B, then either it passes on this message to the TV via channel C, or it loses the message. In the latter case, the subsequent delay triggers the timer T_1 to send a time-out to the RC via channel G.
- If L receives an acknowledgement from the TV via channel E, then either it passes on this acknowledgement to the RC via channel F, or it loses the acknowledgement. In the latter case, the subsequent delay triggers the timer T_1 to send a time-out to the RC via channel G.

Note that the recursive specification E for the BRP is guarded, and that it generates a regular process. In the remainder of this section, for notational convenience, process terms $\langle Y|E \rangle$ are abbreviated to Y. The BRP is expressed by the process term

$$\tau_I(\partial_H(V\|X\|K\|L))$$

where the set H consists of the read and send actions over the internal channels B, C, E, F, G, and H, while the set I consists of the communication actions over these internal channels.

The process term $\tau_I(\partial_H(V\|X\|K\|L))$ exhibits the required external behaviour (see Fig. 6.4), intertwined with non-silent τ-transitions. We proceed to sketch an algebraic derivation of this fact. CFAR does not need to be applied in this derivation, owing to the absence of τ-loops. A detailed verification that the BRP exhibits its required external behaviour is given in [110]. Alternative verifications of the BRP can be found in [1, 78, 119].

The following equations can be derived from $\mathcal{E}_{\text{ACP}}$, commutativity of the merge, and RDP. For notational convenience, process terms are considered modulo associativity of the merge, and $K'(d,b)$, $K'(d,b,\text{last})$, and L' abbreviate $(s_C(d,b)+s_G(to))\cdot K$, $(s_C(d,b,\text{last})+s_G(to))\cdot K$, and $(s_F(\text{ack})+s_G(to))\cdot L$, respectively. The equation below captures the initial state.

$$\partial_H(V\|X\|K\|L) \;=\; \sum_{\ell\in\Lambda} r_A(\ell) \cdot \partial_H(V\|Y(\ell,0,0)\|K\|L)$$

The equation below captures the state in which the RC sends the first datum of a packet, while the TV did not yet receive a datum of this packet.

$$\partial_H(V\|Y(\ell,0,n)\|K\|L) \;=\;$$
$$c_B(\text{head}(\ell),0) \cdot \partial_H(V\|Z(\ell,0,n)\|K'(\text{head}(\ell),0)\|L)$$

The equation below captures the state in which the RC sends some, but not the last, datum of a packet, while the TV already received one or more data elements of this packet.

$$\partial_H(W(b)\|Y(\ell,b',n)\|K\|L) \;=\;$$
$$c_B(\text{head}(\ell),b') \cdot \partial_H(W(b)\|Z(\ell,b',n)\|K'(\text{head}(\ell),b')\|L)$$

The two equations below capture the state in which the RC sends the last datum of a packet. The first equation deals with the case that the TV did not yet receive this datum, while the second equation deals with the case that the TV already received this datum.

$$\partial_H(W(b)\|Y(d,b,n)\|K\|L) = \\ c_B(d,b,last) \cdot \partial_H(W(b)\|Z(d,b,n)\|K'(d,b,last)\|L)$$

$$\partial_H(V\|Y(d,b,n)\|K\|L) = \\ c_B(d,b,last) \cdot \partial_H(V\|Z(d,b,n)\|K'(d,b,last)\|L)$$

The two equations below capture the state in which medium K either passes on or loses the first datum of a packet, while the TV did not yet receive a datum of this packet. The second equation deals with the special case that the counter has reached its maximum value max.

$$\partial_H(V\|Z(\ell,0,n)\|K'(head(\ell),0)\|L) = \\ c_C(head(\ell),0) \cdot s_D(head(\ell)) \cdot c_E(ack) \cdot \partial_H(W(1)\|Z(\ell,0,n)\|K\|L') \\ + c_G(to) \cdot \partial_H(V\|Y(\ell,0,n+1)\|K\|L) \qquad (n < max)$$

$$\partial_H(V\|Z(\ell,0,max)\|K'(head(\ell),0)\|L) = \\ c_C(head(\ell),0) \cdot s_D(head(\ell)) \cdot c_E(ack) \cdot \partial_H(W(1)\|Z(\ell,0,max)\|K\|L') \\ + c_G(to) \cdot s_A(I_{NOK}) \cdot c_H(to) \cdot \partial_H(V\|X\|K\|L)$$

The four equations below capture the state in which medium K either passes on or loses a datum of a packet, while the TV already received some data elements of this packet, but not the datum handled by K. The last two equations deal with the special case that K handles the last datum of a packet. Moreover, the second and fourth equation deal with the special case that the counter has reached its maximum value max.

$$\partial_H(W(b)\|Z(\ell,b,n)\|K'(head(\ell),b)\|L) = \\ c_C(head(\ell),b) \cdot s_D(head(\ell)) \cdot c_E(ack) \cdot \partial_H(W(1-b)\|Z(\ell,b,n)\|K\|L') \\ + c_G(to) \cdot \partial_H(W(b)\|Y(\ell,b,n+1)\|K\|L) \qquad (n < max)$$

$$\partial_H(W(b)\|Z(\ell,b,max)\|K'(head(\ell),b)\|L) = \\ c_C(head(\ell),b) \cdot s_D(head(\ell)) \cdot c_E(ack) \cdot \partial_H(W(1-b)\|Z(\ell,b,max)\|K\|L') \\ + c_G(to) \cdot s_A(I_{NOK}) \cdot c_H(to) \cdot s_D(I_{NOK}) \cdot \partial_H(V\|X\|K\|L)$$

$$\partial_H(W(b)\|Z(d,b,n)\|K'(d,b,last)\|L) = \\ c_C(d,b,last) \cdot s_D(d) \cdot s_D(I_{OK}) \cdot c_E(ack) \cdot \partial_H(V\|Z(d,b,n)\|K\|L') \\ + c_G(to) \cdot \partial_H(W(b)\|Y(\ell,b,n+1)\|K\|L) \qquad (n < max)$$

$$\partial_H(W(b)\|Z(d,b,max)\|K'(d,b,last)\|L) = \\ c_C(d,b,last) \cdot s_D(d) \cdot s_D(I_{OK}) \cdot c_E(ack) \cdot \partial_H(V\|Z(d,b,max)\|K\|L') \\ + c_G(to) \cdot s_A(I_{DK}) \cdot c_H(to) \cdot s_D(I_{NOK}) \cdot \partial_H(V\|X\|K\|L)$$

The four equations below capture the state in which medium K either passes on or loses a datum of a packet, while the TV already received the datum handled by K. The last two equations deal with the special case that K handles the last datum of a packet. Moreover, the second and fourth equation deal with the special case that the counter has reached its maximum value max.

$$\partial_H(W(b)\|Z(\ell,1-b,n)\|K'(head(\ell),1-b)\|L) =$$
$$c_C(head(\ell),1-b) \cdot c_E(ack) \cdot \partial_H(W(b)\|Z(\ell,1-b,n)\|K\|L')$$
$$+ c_G(to) \cdot \partial_H(W(b)\|Y(\ell,1-b,n+1)\|K\|L) \qquad (n < max)$$

$$\partial_H(W(b)\|Z(\ell,1-b,max)\|K'(head(\ell),1-b)\|L) =$$
$$c_C(head(\ell),1-b) \cdot c_E(ack) \cdot \partial_H(W(b)\|Z(\ell,1-b,max)\|K\|L')$$
$$+ c_G(to) \cdot s_A(I_{NOK}) \cdot c_H(to) \cdot s_D(I_{NOK}) \cdot \partial_H(V\|X\|K\|L)$$

$$\partial_H(V\|Z(d,b,n)\|K'(d,b,last)\|L) =$$
$$c_C(d,b,last) \cdot c_E(ack) \cdot \partial_H(V\|Z(d,b,n)\|K\|L')$$
$$+ c_G(to) \cdot \partial_H(V\|Y(d,b,n+1)\|K\|L) \qquad (n < max)$$

$$\partial_H(V\|Z(d,b,max)\|K'(d,b,last)\|L) =$$
$$c_C(d,b,last) \cdot c_E(ack) \cdot \partial_H(V\|Z(d,b,max)\|K\|L')$$
$$+ c_G(to) \cdot s_A(I_{DK}) \cdot c_H(to) \cdot \partial_H(V\|X\|K\|L)$$

The four equations below capture the state in which medium L either passes on or loses an acknowledgement. The last two equations deal with the special case that the acknowledgement concerns the last datum of a packet. Moreover, the second and fourth equation deal with the special case that the counter has reached its maximum value max.

$$\partial_H(W(b)\|Z(\ell,1-b,n)\|K\|L') =$$
$$c_F(ack) \cdot \partial_H(W(b)\|Y(tail(\ell),b,0)\|K\|L)$$
$$+ c_G(to) \cdot \partial_H(W(b)\|Y(\ell,1-b,n+1)\|K\|L) \qquad (n < max)$$

$$\partial_H(W(b)\|Z(\ell,1-b,max)\|K\|L') =$$
$$c_F(ack) \cdot \partial_H(W(b)\|Y(tail(\ell),b,0)\|K\|L)$$
$$+ c_G(to) \cdot s_A(I_{NOK}) \cdot c_H(to) \cdot s_D(I_{NOK}) \cdot \partial_H(V\|X\|K\|L)$$

$$\partial_H(V\|Z(d,b,n)\|K\|L') =$$
$$c_F(ack) \cdot s_A(I_{OK}) \cdot \partial_H(V\|X\|K\|L)$$
$$+ c_G(to) \cdot \partial_H(V\|Y(d,b,n+1)\|K\|L) \qquad (n < max)$$

$$\partial_H(V\|Z(d,b,max)\|K\|L') =$$
$$c_F(ack) \cdot s_A(I_{OK}) \cdot \partial_H(V\|X\|K\|L)$$
$$+ c_G(to) \cdot s_A(I_{DK}) \cdot c_H(to) \cdot \partial_H(V\|X\|K\|L)$$

After application of the abstraction operator τ_I, communication actions over the internal channels B, C, E, F, G, and H are renamed into τ, after which most of these actions can be removed using axiom B1. However, some of the τ's are not truly silent, and the resulting equations capture the external behaviour in Fig. 6.4 intertwined with these non-silent τ-transitions.

Exercise 6.2.1. Give a detailed algebraic derivation of the external behaviour of $\tau_I(\partial_H(V\|X\|K\|L))$.

6.3 Specification and Verification Techniques

Over the last two decades, a large number of specifications and verifications of network protocols by means of process algebra have appeared in the literature. Collections of such verifications can be found in [11, 146]. This section presents a brief overview of standard techniques that are used in these specifications and verifications. For verifications in the specification language μCRL [112] that use one or more of these techniques, see [56, 96, 109, 138, 180]

Expansion. A basic technique in protocol verification is *expansion* [47] of the merge operator. That is, in order to compute the initial transitions of a process term $t_1\|\cdots\|t_n$, it is sufficient to compute the initial transitions of its arguments $t_1,\ldots,t_n$. The verifications of the ABP and of the BRP, which were discussed in Sections 6.1 and 6.2, mainly consisted of such expansions. Moreover, applications of expansion in PAP and ACP can be found in Exercises 3.3.3 and 4.3.6, respectively.

Alphabet Axioms. Baeten, Bergstra, and Klop [21] introduced *alphabet axioms*, to obtain the set of actions that a process term can perform. These axioms allow for instance to eliminate redundant encapsulation and abstraction operators. Namely, if a process term t cannot perform any actions from a set H, then one can derive $\partial_H(t) = t$. Korver and Sellink [137] formulated alphabet axioms in the presence of data parameters.

Language Matching. Language matching was introduced by van Wamel [193] as a method for reducing and labelling traces of actions that are not in a predefined set of traces, called a language. In general this language is defined to consist of the expected traces, and if a process term in the argument of an encapsulation operator behaves as expected, then language matching in combination with the alphabet axioms makes it possible to weed out all labelled traces.

Determinacy and Confluence. Milner [155] propagated the notion of determinacy in process algebra, which enhances the predictability of process behaviour. Let $p \Rightarrow q$ abbreviate that there exists a sequence of transitions $p \xrightarrow{\tau} \cdots \xrightarrow{\tau} q$, and let $p \xRightarrow{a} q$ abbreviate that there exists a sequence of transitions $p \Rightarrow \xrightarrow{a} \Rightarrow q$. A process p is *determinate* modulo branching bisimulation if it satisfies the following two conditions:

1. if $p \xRightarrow{a} q$ and $p \xRightarrow{a} r$, then $q \underline{\leftrightarrow}_b r$;
2. if $p \Rightarrow q$ and $p \Rightarrow r$, then $q \underline{\leftrightarrow}_b r$.

Milner [155] restricted determinacy to a notion of confluence, because the latter notion has better congruence properties. A process p is *confluent* modulo branching bisimulation if it is determinate and satisfies two extra conditions:

3. if $p \stackrel{a}{\Rightarrow} q$ and $p \stackrel{b}{\Rightarrow} r$ with $a \not\equiv b$, then $q \stackrel{b}{\Rightarrow} q'$ and $r \stackrel{a}{\Rightarrow} r'$ with $q' \leftrightarrow_b r'$;

4. if $p \stackrel{a}{\Rightarrow} q$ and $p \Rightarrow r$, then $r \stackrel{a}{\Rightarrow} r'$ with $q \leftrightarrow_b r'$.

Confluence often enables one to substantially reduce the LTS under consideration by identifying states that are branching bisimilar. See [113, 155] for thorough discussions on and examples of the use of confluence in process algebra verifications.

Value Passing. The specifications of the ABP and the BRP make use of *value passing*, meaning that atomic actions and recursion variables carry parameters to pass on data values. Value passing is a standard method in process algebra for including data types in protocol specifications. We mention two examples.

Grid protocols [40] model concurrent systems in a grid-like architecture, based on the design of synchronous concurrent algorithms [185]. A grid protocol consists of a number of data processing units with in- and outgoing ports, via which data is communicated between these units or with the environment. Each unit has a state variable, the value of which is repeatedly updated on the basis of incoming data elements. Grid protocols can be used, among other things, for modelling hardware and for approximating solutions to differential equations.

The π-calculus [158] extends process algebra with explicit port names. The so-called *mobile processes* in the π-calculus are able to communicate port names via ports, thus allowing dynamic reconfiguration of topologies of linked ports. Typically, a term $x(y).t$ sends and a term $\bar{x}(y).t$ reads port name y via port x, after which they proceed as t. (In both expressions, the x is free and the y is bound in the term t.) An elegant structural operational semantics for a subset of the full π-calculus was given in [179]. See [157] for an introduction to the π-calculus.

Invariants. An *invariant* [55] is a dependency relation on data objects in a process algebra specification that holds throughout the states of the process graph that belongs to this specification. This yields a characterisation of the states that are reachable from the root state. Invariants have been used in many process algebra verifications in which data play a prominent role, to facilitate the correctness proof.

Linear Process Operators. The RSP principle (see Section 4.3) can be generalised to a setting with data parameters. The role of guarded linear recursive specifications (see Definition 5.2.1) is then passed on to so-called *linear process operators*, which are symbolic representations of process graphs with explicit data parameters. The principle CL-RSP [55] states that each linear process operator that does not induce infinite sequences of τ-transitions has no more than one solution.

Cones and Focus Points. A *focus point* is a state from which there are no τ-transitions. The *cone* of a focus point is the set of states that can reach the focus point by a series of τ-transitions. Barring infinite sequences of τ-transitions (which may be eliminated by CFAR), each state belongs to a cone. Groote and Springintveld [114] presented a general verification technique for linear process operators that do not induce infinite sequences of τ-transitions. Ideally, this technique enables one to identify the states in a cone with the focus point of this cone, using the following approach. Assume a process graph that belongs to a linear process operator, a process graph without τ-transitions, and a mapping h from states in the first to states in the second process graph, which maps all the states in a cone to the same state. Groote and Springintveld [114] formulated straightforward criteria to ensure that states s and $h(s)$ are branching bisimilar, so that each state s in the first process graph can be identified with the state $h(s)$ in the second process graph.

Conditions and Signals. Data parameters that are used in value-passing, or of which the values are determined by the environment, may influence the execution of processes. This can be modelled using *conditions* ϕ (see [13, 111]), which take as input data parameters, and present as output a boolean value true or false. The expression $\phi :\rightarrow t$ represents the behaviour of the process term t under condition ϕ, meaning that true $:\rightarrow t$ behaves as t and false $:\rightarrow$ t behaves as δ. Bergstra *et al.* [36, 49] showed how to deal with conditions in process algebra when their output domain is a five-valued logic including meaningless, divergence, and choice.

Behaviour can often only get into a specific state if certain external conditions are satisfied. For example, a camp-fire can only be lit if it is not raining, or a train can only traverse a level crossing if its barriers are closed. Such external conditions can be captured by means of *signals* [13, 17], which enable one to eliminate inconsistent states from the process graph that belongs to a process term.

Time. Time often plays an important role in system behaviour. One way to model time in process algebra is by the use of timers, which send time-out messages at random. An example of the use of such timers in process algebra was given in the BRP; see Section 6.2.

If one wants to enforce that actions can only communicate if they are executed at the same moment in time, then time needs to be present as an explicit quantity. This time model depends on a number of design decisions. First, we discuss *discrete* versus *dense* (or *real*) time.

1. Discrete time assumes that time evolves in discrete time steps. In between two time steps, a process behaves as in untimed process algebra. When a time step is made, the process comes to a halt and the value of the clock is increased by one, after which the process continues. Though discrete time does not simulate time as experienced in real life, it does resemble

the timing mechanisms that are used in computers. Baeten and Bergstra [16] introduced an extension of process algebra with discrete time, in which time is modelled by a unary operator σ_{rel}: the process term $\sigma_{rel}(t)$ represents the behaviour of t delayed by one time step.

2. Dense time assumes that time progresses continuously. Its practical use lies for example in modelling biological phenomena. Baeten and Bergstra [14] (see also [92]) extended process algebra with dense time by supplying atomic actions with time stamps: the action a with as time stamp the positive real number r represents the action a that is executed at time r.

Time stamps of timed actions can relate either to an *absolute* or to a *relative* clock. In absolute time, a timed action $a(r)$ executes a at time r. For example, the process that executes action a at the start of every time unit can be described by

$$a(1) \cdot a(2) \cdot a(3) \cdot \cdots$$

or by $\langle X(1)|E \rangle$, where E consists of the following recursive equations, for $k \in \mathbb{N}$:

$$X(k) = a(k) \cdot X(k+1).$$

In relative time, a timed action $a[r]$ executes a exactly r time units after the previous action was executed. For example, the process that executes action a at the start of every time unit can be described by

$$a[1] \cdot a[1] \cdot a[1] \cdot \cdots$$

or by $\langle Y|F \rangle$, where F consists of the recursive equation

$$Y = a[1] \cdot Y.$$

We proceed to discuss the modelling of some special constants in process algebra with relative dense time. The timed deadlock behaves as follows. The expression $\delta[r]$ represents a deadlock at relative time r; that is, it can idle for r time units after the previous action was executed, and then it gets stuck. For example, the process term $a[1] + \delta[2]$ may get into a deadlock at time 2, if it idles beyond time 1 without executing the action a. On the other hand, the process term $a[2] + \delta[1]$ cannot get into a deadlock, because it can idle beyond time 1 to execute the action a at time 2.

The introduction of the silent step in time is more complicated. The intuition behind branching bisimulation is that a τ-transition is truly silent if and only if it does not lose possible behaviours. The same intuition in the timed case gives rise to quite a different mathematical interpretation than in the untimed case. We give two examples in relative time.

- In the untimed setting, $\tau(a + b) + a \underline{\leftrightarrow}_b a + b$. However,

$$\tau[1] \cdot (a[1] + b[1]) + a[2] \not\leftrightarrow_b a[2] + b[2].$$

Not executing the τ at time 1 in the process term on the left means a decision that the a, and not the b, will be executed at time 2.

- In the untimed setting, $\tau a + b \not\leftrightarrow_b a + \tau b$. However,

$$\tau[1] \cdot a[1] + b[2] \leftrightarrow_b a[2] + \tau[1] \cdot b[1].$$

In both process terms it is decided at time 1 whether the a or the b will be executed at time 2.

Timed branching bisimulation equivalence was studied in [89, 133, 134].

In protocols that are studied in the literature, the use of time can be more complicated than was discussed here. For instance, each separate component of a concurrent process may have its own local clock, where these local clocks all refer (either precisely or roughly) to a global clock. Such protocols may require a more advanced time model.

Probabilities. Probabilities can be of importance in applications of process algebra, because actions are not alway executed with the same probability. One can give weights to atomic actions, to express the chance that such an action is executed. For example, the process term $(a, 0.25) + (b, 0.75)$ has 25% chance of executing action a, and 75% chance of executing action b. In general, in each state of a process, the probabilities of executing the possible actions in that state should add up to 1 (that is, to 100%), or in any case not exceed 1. See [24, 187] for expositions on probabilistic process algebra, and [141] for a notion of probabilistic bisimulation. A process algebra with random clocks and stochastic time behaviour is described in [77].

6.4 Tools

As the case-studies that are tackled using process algebra are becoming more and more complicated, tool support for the analysis of concurrent systems is becoming increasingly important. In recent years, a wide range of tool environments have been developed that are based on process algebra, modal and temporal logics (see Section B.6), and general proof techniques. Such tool environments comprise standard features that are familiar from the world of programming languages, such as a type-checker and a compiler. Furthermore, they incorporate features that aim specifically at the analysis of process terms and finite-state process graphs.

1. A *graph generator* produces the process graph that belongs to a process term.
2. An *equivalence checker* verifies whether two states in a process graph are equivalent with respect to some process equivalence, such as bisimulation or rooted branching bisimulation.

3. A *minimiser* reduces the number of states in a process graph. Such a minimiser can identify states that are equivalent modulo some process equivalence, or apply so-called *partial-order reduction* to eliminate redundant states that are the result of interleaving unrelated events.

4. A *simulator* runs a random trace in a process graph, to test it, for instance, on the presence of deadlocks.

5. A *term rewriter* reduces process or data terms to normal form, with respect to some term rewriting system.

6. A *model checker* verifies whether a state in a process graph satisfies a requirement formulated in some modal or temporal logic.

7. A *theorem prover* is geared to automatically derive mathematical theorems from a set of assumptions and previously proven results.

We proceed to present an (admittedly incomplete) overview of existing specification languages and tool environments that support the verification of concurrent systems.

- *LOTOS* (Language of Temporal Ordering Specifications) [60] is a widely used specification language based on process algebra. It is combined with ACT ONE, being an algebraic specification language for data types. A number of tools have been based on LOTOS, some of which are discussed in some detail below.

- *CADP* (Cæsar/Aldébaran Development Package) [88] is a French verification tool box for LOTOS specifications, which supports the use of data types specified in ACT ONE. Cæsar generates the process graph belonging to a LOTOS specification, and supports simulation. Aldébaran performs equivalence checking and minimisation with respect to such process graphs modulo a range of process equivalences. XTL offers facilities for model checking formulas in modal and temporal logics such as HML and ACTL (see Section B.6).

- *XEludo* [117] from Canada provides facilities for the simulation of LOTOS specifications, during which the user is prompted for data input when necessary. It supports model checking of CTL formulas via a stand alone tool LMC.

- μ*CRL* (Micro Common Representation Language) [112] is a Dutch specification language that targets the specification and manipulation of data in process verification. Its tool set, which is based on linear process operators, includes simulation and term rewriting facilities, and the generated process graphs are suitable as input to the CADP tool box.

- *PSF* (Process Specification Formalism) [145] is a Dutch tool kit based on ACP, in which data can be specified using the modular approach propagated in [39]. It supports equivalence checking, simulation, and term rewriting.

- The *Concurrency Workbench Edinburgh* [74] is a tool environment for the analysis of concurrent systems, based on CCS and timed CCS [160]. The *Concurrency Workbench North Carolina* [75] has the same ancestor as its sibling in Edinburgh, but is now under separate development. Several front-ends allow the analysis of specifications in untimed and timed CCS, CSP, and LOTOS. These tool environments incorporate equivalence checking and reduction with respect to a range of process equivalences, simulation, and model checking formulas in the modal μ-calculus. The *Concurrency Factory* [73] can be viewed as a next generation of the latter Concurrency Workbench. It supports basic data types, and minimises the process graph under consideration by partial-order reduction.

- *FC2Tools* [61] is a French verification tool kit that can cope with graphical representations of automata and with CCS and LOTOS expressions. It supports equivalence checking, minimisation modulo process equivalences, and on-the-fly model checking. On-the-fly means that a formula is checked while the process graph is under construction.

- *Esterel* [52] is a French synchronous reactive programming language, which supports the algebraic specification of data types, and has been supplied with a structural operational semantics [53]. *Xeve* is a tool environment for the verification of Esterel programs, modelled as process graphs, which includes minimisation modulo process equivalences, and model checking LTL formulas.

- *FDR* (Failures-Divergence Refinement) is a commercial British tool environment, based on value-passing CSP. There is a simulator *ProBE* for CSP process expressions, it allows model checking, and has extensive debugging facilities.

- The Australian hardware description and verification language *XCircal* is based on the process algebra Circal [148], featuring so-called 'multi-point' communication and a distinction between deterministic and non-deterministic alternative composition. The *Circal System*, an implementation of XCircal that incorporates simulation, equivalence checking, and a notion of discrete time, is being used in the verification of digital hardware.

- *VERSA* (Verification, Execution, and Rewrite System of ACSR) [69] from the USA is based on the dense-time process algebra ACSR [63] with resource-specific delays and priority arbitration. The tool set XVERSA supports simulation, term rewriting, equivalence checking, and model checking.

- *SMV* (Symbolic Model Verifier) [147] is an automated model checker for CTL formulas from the USA. It was one of the first to represent process graphs by so-called binary decision diagrams [65], which provide a compact notation for boolean formulas. Owing to this representation, model checking has been performed with respect to process graphs consisting of more than 10^{20} states; see [72]. SMV also has a diagnostic facility that produces a counter-example when a CTL formula is found to be false. NuSMV [68] is a reimplementation and extension of SMV from Italy.

- *Spin* [128], developed in the USA, allows simulation and model checking of LTL formulas. Model checking is performed on-the-fly and using partial-order reduction. Moreover, model checking can be done in a conventional exhaustive search through the process graph, or, when this graph is too large, with an efficient approximation method. Spin supports the specification of basic data types.

- *XMC* [176] is a model checker from the USA for value-passing CCS, to calculate the validity of formulas in the modal μ-calculus. It has been implemented in logic programming, using SLD resolution and so-called tabled resolution.
- *Murφ* [83] is a model checker for LTL formulas from the USA, based on explicit state enumeration. While constructing the process graph under consideration, multiple construction of the same state is avoided. Symmetry properties of process graphs are used to further reduce the state space. Murφ supports the use of basic data types.
- *COSPAN* (COordinated SPecification ANalysis), [118] checks on so-called language containment of ω-automata, to see whether each trace of actions that can be performed by the implementation can also be performed by the specification. It uses either explicit state enumeration or an algorithm based on binary decision diagrams. COSPAN supports some basic data types and provides an error-tracing facility.
- *STeP* (Stanford Temporal Prover) [144] combines theorem proving techniques with model checking of LTL formulas with respect to systems that can be parametrised over infinite data domains.
- *UPPAAL* (Uppsala Aalborg) [34], named after the two sites where it was constructed, is a tool suite for the verification of dense-time systems, which allows one to graphically specify networks of timed automata [5]. UPPAAL can perform a reachability analysis, and it supports simulation and diagnostic error-trace reports.
- *Kronos* [79] from France supports minimisation of timed automata modulo process equivalences, and model-checking formulas in TCTL [4], which is a dense-time extension of CTL.
- *SGM* (State Graph Manipulators) [129] from Taiwan targets the reduction of timed automata, together with model checking of TCTL formulas.

Finally, some popular theorem provers are PVS [163], HOL [106], Isabelle [167], and Nqthm [62]. For more information and internet links, see [76, 95].

7. Extensions

Baeten, Bergstra, and Klop [19] proved that every computable process (see [165]) can be specified by means of a process term in ACP_τ with guarded recursion. Namely, it is possible to specify a Turing machine [188] in this algebra. In spite of the expressive power of ACP_τ with guarded recursion, it is important to realise that there is no need to restrict to this framework. Often a protocol can be specified more easily with the help of some auxiliary operator, to express a particular feature of the protocol in an elegant fashion. In this case, one must formulate transition rules for the new operator, check that they are within the formats for conservative extension and congruence, and come up with a sound axiomatisation, which ideally is also complete.

In the next sections we present examples of auxiliary operators, which do not increase the expressivity of process algebra, but which have proven to be useful for the specification of system behaviour.

7.1 Renaming

It can be convenient to rename atomic actions. From a theoretical point of view, such a renaming construct is interesting because it allows one to derive CFAR from the more elegant (but weaker) axiom KFAR; see [189].

The unary *renaming operator* ρ_f assumes a renaming function $f : A \to A$. The process graph of a process term $\rho_f(t)$ is obtained by renaming all labels a of transitions in the process graph of t into $f(a)$. This general renaming concept was introduced by Milner [151]. The transition rules for renaming operators are as follows, where f is extended to $A \cup \{\tau\}$ by defining $f(\tau) \overset{\Delta}{=} \tau$:

$$\frac{x \overset{v}{\to} \checkmark}{\rho_f(x) \overset{f(v)}{\to} \checkmark} \qquad \frac{x \overset{v}{\to} x'}{\rho_f(x) \overset{f(v)}{\to} \rho_f(x')}$$

The variables x and x' range over process terms, while v ranges over $A \cup \{\tau\}$.

Theorem 7.1.1. *ACP_τ with guarded linear recursion and renaming operators is a conservative extension of ACP_τ with guarded linear recursion.*

Proof. The sources of the transition rules for the renaming operator contain the fresh function symbol ρ_f. Since furthermore the transition rules of ACP_τ

with guarded linear recursion are source-dependent, the extension of this algebra with renaming operators is conservative; see Theorem B.5.1. □

Theorem 7.1.2. *Rooted branching bisimulation equivalence is a congruence with respect to ACP_τ with guarded linear recursion and renaming operators.*

Proof. As in the proof of Theorem 5.2.2, the transition rules of ACP_τ with guarded linear recursion and the renaming operator can be brought into RBB cool format, by incorporating the successful termination predicate $\downarrow$. This implies that the rooted branching bisimulation equivalence induced by this TSS is a congruence; see Theorem B.4.1. □

Table 7.1 presents axioms for the renaming operators. The variables x and y range over process terms, while v ranges over $A \cup \{\tau\}$.

Table 7.1. Axioms for renaming

RN1	$\rho_f(v) = f(v)$
RN2	$\rho_f(\delta) = \delta$
RN3	$\rho_f(x + y) = \rho_f(x) + \rho_f(y)$
RN4	$\rho_f(x \cdot y) = \rho_f(x) \cdot \rho_f(y)$

Theorem 7.1.3. $\mathcal{E}_{ACP_\tau} + RDP, RSP, CFAR + RN1\text{-}4$ *is sound for ACP_τ with guarded linear recursion and renaming operators, modulo rooted branching bisimulation equivalence.*

Proof. Since rooted branching bisimulation is both an equivalence and a congruence, we only need to check that if $s = t$ is an axiom and σ a closed substitution that maps the variables in s and t to process terms, then $\sigma(s) \underline{\leftrightarrow}_{rb} \sigma(t)$. Here, we only provide some intuition for soundness of the axioms in Table 7.1:

- RN1,2 are the defining equations for the renaming operator ρ_f: RN1 says that it renames atomic actions a into $f(a)$, while RN2 says that it leaves the deadlock δ unchanged;
- RN3,4 say that in $\rho_f(t)$, the labels of all transitions of t are renamed by means of the mapping f.

These intuitions can be made rigorous by means of explicit rooted branching bisimulation relations between the left- and right-hand sides of closed instantiations of RN1-4. □

Theorem 7.1.4. $\mathcal{E}_{ACP_\tau} + RDP, RSP, CFAR + RN1\text{-}4$ *is complete for ACP_τ with guarded linear recursion and renaming operators, modulo rooted branching bisimulation equivalence.*

Proof. It suffices to prove that each process term t in ACP_τ with guarded linear recursion and renaming operators is provably equal to a process term $\langle X|E \rangle$ with E a guarded linear recursive specification. Namely, then the desired completeness result follows from the fact that if $\langle X_1|E_1 \rangle \underleftrightarrow{}_{rb} \langle Y_1|E_2 \rangle$ for guarded linear recursive specifications E_1 and E_2, then $\langle X_1|E_1 \rangle = \langle Y_1|E_2 \rangle$ can be derived from $\mathcal{E}_{\text{ACP}} + \text{B1}, 2 + \text{RDP}, \text{RSP}$; see the proof of Theorem 5.3.2.

We apply structural induction with respect to process term t. In comparison to the completeness proof of Theorem 5.6.2, the only new case (where RN1-4 are needed) is when $t \equiv \rho_f(s)$. By induction we may assume that $s = \langle X_1|E \rangle$ with E a guarded linear recursive specification, so $t = \rho_f(\langle X_1|E \rangle)$. Let E consist of linear recursive equations

$$X_i = a_{i1}X_{i1} + \cdots + a_{ik_i}X_{ik_i} + b_{i1} + \cdots + b_{i\ell_i}$$

for $i \in \{1, \ldots, n\}$. The recursive specification F is defined to consist of the linear recursive equations

$$Y_i = f(a_{i1})Y_{i1} + \cdots + f(a_{ik_i})Y_{ik_i} + f(b_{i1}) + \cdots + f(b_{i\ell_i})$$

for $i \in \{1, \ldots, n\}$. Since E is guarded, it follows that F is also guarded. (This observation uses in an essential way that $f(a) \not\equiv \tau$ for $a \in A$.)

$$\rho_f(\langle X_i|E \rangle)$$
$$\overset{\text{RDP}}{=} \rho_f(a_{i1}\langle X_{i1}|E \rangle + \cdots + a_{ik_i}\langle X_{ik_i}|E \rangle + b_{i1} + \cdots + b_{i\ell_i})$$
$$\overset{\text{RN1-4}}{=} \rho_f(a_{i1}) \cdot \rho_f(\langle X_{i1}|E \rangle) + \cdots + \rho_f(a_{ik_i}) \cdot \rho_f(\langle X_{ik_i}|E \rangle)$$
$$\qquad + \rho_f(b_{i1}) + \cdots + \rho_f(b_{i\ell_i}).$$

Hence, replacing Y_i by $\rho_f(\langle X_i|E \rangle)$ for $i \in \{1, \ldots, n\}$ is a solution for F. So by RSP, $\rho_f(\langle X_1|E \rangle) = \langle Y_1|F \rangle$. $\square$

Exercise 7.1.1. Assume a renaming function $f : A \to A$ with $f(a) \overset{\Delta}{=} c$ and $f(b) \overset{\Delta}{=} c$. Derive $\rho_f(\langle X \mid X=aX+bX \rangle) = \langle Y \mid Y=cY \rangle$ from the axioms.

Exercise 7.1.2. Assume renaming functions $f : A \to A$ and $g : A \to A$. Derive the equation $\rho_{g \circ f}(t) = \rho_g(\rho_f(t))$ from the axioms for process terms t in ACP_τ with renaming.

7.2 State Operator

In Chapter 4 it was shown that one way to describe a regular process is by means of a linear recursive specification. Each state in the protocol is assigned its own recursion variable, and the linear recursive specification expresses the transitions between the different states. This section describes an alternative method to capture the states of a regular process, by means of a so-called state operator, introduced by Baeten and Bergstra [12, 15].

Let S denote a finite set of states. We assume that the visible behaviour of an action a depends on the state in which it is executed, and that such an execution causes the transposition to a new state. This is expressed by two mappings:

$$action : S \times A \to A$$
$$effect : S \times A \to S.$$

Intuitively, $action(s,a)$ represents the visible behaviour of action a in state s, while $effect(s,a)$ represents the state that results if action a is executed in state s. The *state operator* $\lambda_s(t)$ denotes process term t in state s. The transition rules for the state operator are as follows, where *action* and *effect* are extended to $A \cup \{\tau\}$ by defining $action(s,\tau) \stackrel{\Delta}{=} \tau$ and $effect(s,\tau) \stackrel{\Delta}{=} s$:

$$\frac{x \stackrel{v}{\to} \sqrt{}}{\lambda_s(x) \stackrel{action(s,v)}{\to} \sqrt{}} \qquad \frac{x \stackrel{v}{\to} x'}{\lambda_s(x) \stackrel{action(s,v)}{\to} \lambda_{effect(s,v)}(x')}$$

The variables x and x' range over process terms, while v ranges over $A \cup \{\tau\}$.

Theorem 7.2.1. *ACP_τ with guarded linear recursion and the state operator is a conservative extension of ACP_τ with guarded linear recursion.*

Proof. The sources of the transition rules for the state operator contain the fresh function symbol λ_s. Since furthermore the transition rules of ACP_τ with guarded linear recursion are source-dependent, the extension of this algebra with the state operator is conservative; see Theorem B.5.1. □

Theorem 7.2.2. *Rooted branching bisimulation equivalence is a congruence with respect to ACP_τ with guarded linear recursion and the state operator.*

Proof. As in the proof of Theorem 5.2.2, the transition rules of ACP_τ with guarded linear recursion and the state operator can be brought into RBB cool format, by incorporating the successful termination predicate $\downarrow$. This implies that the rooted branching bisimulation equivalence induced by this TSS is a congruence; see Theorem B.4.1. □

Table 7.2 presents axioms for the state operator. The variables x and y range over process terms, v ranges over $A \cup \{\tau\}$, and s ranges over the set S of states.

Theorem 7.2.3. *$\mathcal{E}_{ACP_\tau} + RDP, RSP, CFAR + SO1\text{-}4$ is sound for ACP_τ with guarded linear recursion and the state operator, modulo rooted branching bisimulation equivalence.*

Proof. Since rooted branching bisimulation is both an equivalence and a congruence, we only need to check that if $s = t$ is an axiom and σ a closed substitution that maps the variables in s and t to process terms, then $\sigma(s) \underline{\leftrightarrow}_{rb} \sigma(t)$. Here, we only provide some intuition for soundness of the axioms in Table 7.2:

Table 7.2. Axioms for the state operator

SO1	$\lambda_s(v) = action(s, v)$
SO2	$\lambda_s(\delta) = \delta$
SO3	$\lambda_s(x + y) = \lambda_s(x) + \lambda_s(y)$
SO4	$\lambda_s(v \cdot y) = action(s, v) \cdot \lambda_{\mathit{effect}(s,v)}(y)$

- SO1,4 say that $\lambda_s(t)$ can execute the visible behaviour $action(s, a)$ of an initial a-transition of t in state s; if the a-transition is a successful termination, then the $action(s, a)$-transition is also a successful termination, while if the a-transition is not a successful termination, then the $action(s, a)$-transition results in the state $effect(s, a)$;
- SO2 says that $\lambda_s(\delta)$ does not exhibit any behaviour;
- SO3 says that in a term $\lambda_s(t + u)$, a choice for an initial transition from t or u is a choice for $\lambda_s(t)$ or $\lambda_s(u)$.

These intuitions can be made rigorous by means of explicit rooted branching bisimulation relations between the left- and right-hand sides of closed instantiations of SO1-4. □

Theorem 7.2.4. $\mathcal{E}_{\mathrm{ACP}_\tau} + \mathrm{RDP}, \mathrm{RSP}, \mathrm{CFAR} + \mathrm{SO1\text{-}4}$ *is complete for* ACP_τ *with guarded linear recursion and the state operator, modulo rooted branching bisimulation equivalence.*

Proof. It suffices to prove that each process term t in ACP_τ with guarded linear recursion and the state operator is provably equal to a process term $\langle X|E \rangle$ with E a guarded linear recursive specification. Namely, then the desired completeness result follows from the fact that if $\langle X_1|E_1 \rangle \underset{rb}{\leftrightarrow} \langle Y_1|E_2 \rangle$ for guarded linear recursive specifications E_1 and E_2, then $\langle X_1|E_1 \rangle = \langle Y_1|E_2 \rangle$ can be derived from $\mathcal{E}_{\mathrm{ACP}} + \mathrm{B1}, 2 + \mathrm{RDP}, \mathrm{RSP}$; see the proof of Theorem 5.3.2.

We apply structural induction with respect to the size of t. In comparison to the completeness proof of Theorem 5.6.2, the only new case (where SO1-4 are needed) is when $t \equiv \lambda_{s_0}(u)$. By induction we may assume that $u = \langle X_1|E \rangle$ with E a guarded linear recursive specification, so $t = \lambda_{s_0}(\langle X_1|E \rangle)$. Let E consist of linear recursive equations

$$X_i = a_{i1}X_{i1} + \cdots + a_{ik_i}X_{ik_i} + b_{i1} + \cdots + b_{i\ell_i}$$

for $i \in \{1, \ldots, n\}$. The recursive specification F is defined to consist of the linear recursive equations

$$\begin{aligned} Y_i(s) = {} & action(s, a_{i1}) \cdot Y_{i1}(effect(s, a_{i1})) \\ & + \cdots + action(s, a_{ik_i}) \cdot Y_{ik_i}(effect(s, a_{ik_i})) \\ & + action(s, b_{i1}) + \cdots + action(s, b_{i\ell_i}) \end{aligned}$$

for $i \in \{1, \ldots, n\}$ and $s \in S$. Since E is guarded, F is also guarded. (This observation uses in an essential way that $action(s, a) \not\equiv \tau$ for $a \in A$.)

$$\lambda_s(\langle X_i | E\rangle)$$
$$\overset{\text{RDP}}{=} \lambda_s(a_{i1}\langle X_{i1}|E\rangle + \cdots + a_{ik_i}\langle X_{ik_i}|E\rangle + b_{i1} + \cdots + b_{i\ell_i})$$
$$\overset{\text{SO1-4}}{=} action(s, a_{i1}) \cdot \lambda_{effect(s, a_{i1})}(\langle X_{i1}|E\rangle)$$
$$+ \cdots + action(s, a_{ik_i}) \cdot \lambda_{effect(s, a_{ik_i})}(\langle X_{ik_i}|E\rangle)$$
$$+ action(s, b_{i1}) + \cdots + action(s, b_{i\ell_i}).$$

Hence, replacing $Y_i(s)$ by $\lambda_s(\langle X_i|E\rangle)$ for $i \in \{1, \ldots, n\}$ and $s \in S$ is a solution for F. So by RSP, $\lambda_{s_0}(\langle X_1|E\rangle) = \langle Y_1(s_0)|F\rangle$. $\square$

The following example of the use of the state operator originates from [12] (see also [28]).

Example 7.2.1. Consider a light that can be switched on and off at two different locations, called X and Y. Both switches can be in two different positions 0 and 1, and the set of states is $\{\langle i, j\rangle \mid i, j \in \{0, 1\}\}$, where $\langle i, j\rangle$ represents the state in which switch X is in position i and switch Y is in position j. The light is on if X and Y are in the same position 0 or 1, and otherwise the light is off. Initially, switch X is in position 0 and switch Y is in position 1, so the light is off. This situation is depicted in Fig. 7.1.

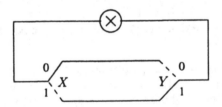

Fig. 7.1. A light switch

The set of atomic actions consists of a, b, *on*, and *off*, where a and b represent flipping the switches at locations X and Y, respectively, and *on* and *off* represent turning the light on and off. All communications between atomic actions result to δ. The recursive equations for the two switches are:

$$X = aX$$
$$Y = bY.$$

In order to specify the system in Fig. 7.1 using a state operator, we need to define the mappings *action* and *effect*. These definitions are limited to the atomic actions a and b; the definitions for *on* and *off* are not of interest, because these atomic actions do not occur in the recursive equations for X and Y. Let i and j range over $\{0, 1\}$:

$$action(\langle i,i\rangle,a) \quad \triangleq off \qquad effect(\langle i,j\rangle,a) \triangleq \langle 1-i,j\rangle$$
$$action(\langle i,1-i\rangle,a) \triangleq on$$
$$action(\langle i,i\rangle,b) \quad \triangleq off \qquad effect(\langle i,j\rangle,b) \triangleq \langle i,1-j\rangle$$
$$action(\langle i,1-i\rangle,b) \triangleq on$$

The definition of *action* reflects that in a state $\langle i,i\rangle$ the light is on, so that an action a or b turns the light off; vice versa, in a state $\langle i,1-i\rangle$ the light is off, so that an action a or b turns the light on. The definition of *effect* reflects that in a state $\langle i,j\rangle$, action a flips the i, while action b flips the j. The initial situation of the system in Fig. 7.1 is captured by the process term

$$\lambda_{\langle 0,1\rangle}(\langle X \mid X{=}aX\rangle \| \langle Y \mid Y{=}bY\rangle).$$

We abbreviate $\langle X \mid X{=}aX\rangle \| \langle Y \mid Y{=}bY\rangle$ to t, and proceed to show that $\lambda_{\langle 0,1\rangle}(t)$ displays the expected external behaviour; i.e., $\lambda_{\langle 0,1\rangle}(t) = on \cdot off \cdot \lambda_{\langle 0,1\rangle}(t)$. Since $\gamma(a,b) \equiv \delta$, we can derive from $\mathcal{E}_{\mathrm{ACP}}{+}\mathrm{RDP}, \mathrm{RSP}$ the equation $t = at + bt$ (cf. the fifth equation in Exercise 4.3.3). Thus,

$$\lambda_{\langle 0,1\rangle}(t)$$
$$= \lambda_{\langle 0,1\rangle}(at + bt)$$
$$\overset{SO3}{=} \lambda_{\langle 0,1\rangle}(at) + \lambda_{\langle 0,1\rangle}(bt)$$
$$\overset{SO4}{=} action(\langle 0,1\rangle,a) \cdot \lambda_{effect(\langle 0,1\rangle,a)}(t) + action(\langle 0,1\rangle,b) \cdot \lambda_{effect(\langle 0,1\rangle,b)}(t)$$
$$\equiv on \cdot \lambda_{\langle 1,1\rangle}(t) + on \cdot \lambda_{\langle 0,0\rangle}(t).$$

In a similar fashion we can derive three more equations:

$$\lambda_{\langle 1,0\rangle}(t) = on \cdot \lambda_{\langle 1,1\rangle}(t) + on \cdot \lambda_{\langle 0,0\rangle}(t)$$
$$\lambda_{\langle 1,1\rangle}(t) = off \cdot \lambda_{\langle 0,1\rangle}(t) + off \cdot \lambda_{\langle 1,0\rangle}(t)$$
$$\lambda_{\langle 0,0\rangle}(t) = off \cdot \lambda_{\langle 0,1\rangle}(t) + off \cdot \lambda_{\langle 1,0\rangle}(t).$$

Let the guarded linear recursive specification E be defined by

$$Z_1 = on \cdot Z_3 + on \cdot Z_4$$
$$Z_2 = on \cdot Z_3 + on \cdot Z_4$$
$$Z_3 = off \cdot Z_1 + off \cdot Z_2$$
$$Z_4 = off \cdot Z_1 + off \cdot Z_2.$$

According to the four derivations above, a solution for E is

$$Z_1 := \lambda_{\langle 0,1\rangle}(t)$$
$$Z_2 := \lambda_{\langle 1,0\rangle}(t)$$
$$Z_3 := \lambda_{\langle 1,1\rangle}(t)$$
$$Z_4 := \lambda_{\langle 0,0\rangle}(t).$$

So by RSP,

$$\lambda_{\langle 0,1\rangle}(t) = \langle Z_1 \mid E\rangle. \tag{7.1}$$

It is easy to see, using RDP and A3, that a second solution for E is

$$Z_1 := \langle W \mid W = on \cdot off \cdot W \rangle$$
$$Z_2 := \langle W \mid W = on \cdot off \cdot W \rangle$$
$$Z_3 := off \cdot \langle W \mid W = on \cdot off \cdot W \rangle$$
$$Z_4 := off \cdot \langle W \mid W = on \cdot off \cdot W \rangle.$$

So by RSP,

$$\langle W \mid W = on \cdot off \cdot W \rangle = \langle Z_1 | E \rangle. \tag{7.2}$$

Equations (7.1) and (7.2) together yield $\lambda_{\langle 0,1 \rangle}(t) = \langle W \mid W = on \cdot off \cdot W \rangle$. So using RDP it follows that

$$\lambda_{\langle 0,1 \rangle}(t) = on \cdot off \cdot \lambda_{\langle 0,1 \rangle}(t).$$

Exercise 7.2.1. Prove that in Example 7.2.1, $\lambda_{\langle 0,0 \rangle}(t) = off \cdot on \cdot \lambda_{\langle 0,0 \rangle}(t)$.

Exercise 7.2.2. Let $A \triangleq \{push, on, off\}$ and $S \triangleq \{0, 1\}$, where intuitively state 0 represents that some machine is off, and state 1 that this same machine is on. Use the state operator to specify a button, such that pushing this button alternately turns the machine on and off. That is, define mappings $action : S \times A \rightarrow A$ and $effect : S \times A \rightarrow S$ such that

$$\lambda_0(\langle X \mid X = push \cdot X \rangle) = on \cdot off \cdot \lambda_0(\langle X \mid X = push \cdot X \rangle).$$

Derive the equation above from the axioms for the state operator, using your definitions for the mappings $action$ and $state$.

Exercise 7.2.3. Let $A \triangleq \{a, b, c\}$, and suppose it would be allowed to have an infinite set of states $\{s_k \mid k \in \mathbb{N}\}$. Give an example of mappings $action$ and $state$ such that the process graph belonging to $\lambda_{s_0}(\langle X \mid X = cX \rangle)$ is not regular.

Exercise 7.2.4. Consider a buffer that can be in two states: in state 1 the buffer is active so that it can read data from a finite, non-empty set Δ, while in state 0 the buffer is inactive. Initially, the buffer is inactive. The atomic action $switch$ represents turning the switch of the buffer, on and off represent turning the buffer on and off, respectively, $read(d)$ for $d \in \Delta$ represents that the buffer receives datum d, and $lost$ represents that the buffer fails to receive such a datum. All communications between atomic actions result to δ. The recursive equations for the switch and for the active buffer are:

$$X = switch \cdot X$$
$$Y = \sum_{d \in \Delta} read(d) \cdot Y.$$

The mappings $action$ and $effect$ are defined as follows (the atomic actions on and off are omitted from these definitions, because they are not of interest):

$$action(0, read(d)) \triangleq lost \qquad\qquad effect(0, read(d)) \triangleq 0$$
$$action(1, read(d)) \triangleq read(d) \qquad effect(1, read(d)) \triangleq 1$$
$$action(0, switch) \triangleq on \qquad\qquad effect(0, switch) \triangleq 1$$
$$action(1, switch) \triangleq off \qquad\qquad effect(1, switch) \triangleq 0$$

Let t abbreviate $\langle X \mid X = switch \cdot X \rangle \| \langle Y \mid Y = \sum_{d \in \Delta} read(d) \cdot Y \rangle$. Prove that $V := \lambda_0(t)$ and $W := \lambda_1(t)$ is a solution for the recursive specification

$$V = lost \cdot V + on \cdot W$$
$$W = off \cdot V + \sum_{d \in \Delta} read(d) \cdot W.$$

7.3 Priorities

In system behaviour it is often the case that an action b is more urgent than some other action a. This means that action a is only executed if it is not possible to execute action b at the same time. This situation can be modelled using the unary *priority operator* Θ, introduced by Baeten, Bergstra, and Klop [18]. This operator assumes a partial order $<$ on $A \cup \{\tau\}$, which is required to be anti-reflexive (i.e., $a < a$ never holds) and transitive (i.e., if $a < b$ and $b < c$, then $a < c$). Intuitively, the process graph of $\Theta(t)$ is obtained by eliminating all transitions $s \xrightarrow{a} s'$ from the process graph of t for which there is a transition $s \xrightarrow{b} s''$ with $a < b$. This is captured by the following transition rules for the priority operator:

$$\frac{x \xrightarrow{v} \surd \quad x \not\xrightarrow{w} \text{ for } v < w}{\Theta(x) \xrightarrow{v} \surd} \qquad\qquad \frac{x \xrightarrow{v} x' \quad x \not\xrightarrow{w} \text{ for } v < w}{\Theta(x) \xrightarrow{v} \Theta(x')}$$

In these transition rules, the negative premise $x \not\xrightarrow{w}$ (see Section B.2) denotes that there does not exist a transition $x \xrightarrow{w} x''$ for any process term x'', and that the transition $x \xrightarrow{w} \surd$ does not hold either.

Recall that the merge could only be axiomatised completely by the introduction of two auxiliary operators left merge and communication merge; see Section 3.2. Similarly, in order to completely axiomatise the priority operator, we use an auxiliary unless operator $x \triangleleft y$. Intuitively, the process graph of $s \triangleleft t$ is obtained by eliminating all initial transitions $s \xrightarrow{a} s'$ from the process graph of s for which there is a transition $t \xrightarrow{b} t'$ with $a < b$. This is captured by the following transition rules for the unless operator:

$$\frac{x \xrightarrow{v} \surd \quad y \not\xrightarrow{w} \text{ for } v < w}{x \triangleleft y \xrightarrow{v} \surd} \qquad\qquad \frac{x \xrightarrow{v} x' \quad y \not\xrightarrow{w} \text{ for } v < w}{x \triangleleft y \xrightarrow{v} x'}$$

The variables x, x', and y in the transition rules for the priority and unless operators range over process terms, while v and w range over $A \cup \{\tau\}$.

The TSS of ACP_τ with guarded linear recursion and the priority and unless operators is positive after reduction (see Definition B.2.4). This can be seen by giving a stratification for this TSS (see Definition B.2.5), which consists of a weight function on transitions such that for each closed substitution instance of a transition rule, the positive premises are smaller or equal than the conclusion, and the negative premises are strictly smaller than the conclusion; see Theorem B.2.1. Since the TSS is positive after reduction, its generated LTS consists of the true transitions in its three-valued stable model (see Definition B.2.2).

Exercise 7.3.1. Give a stratification for the TSS of ACP_τ with guarded linear recursion and the priority and unless operators.

Theorem 7.3.1. *ACP_τ with guarded linear recursion and the priority and unless operators is a conservative extension of ACP_τ with guarded linear recursion.*

Proof. The sources of the transition rules for the priority operator contain the fresh function symbol Θ, and the sources of the transition rules for the unless operator contain the fresh function symbol $\triangleleft$. Since furthermore the transition rules of ACP_τ with guarded linear recursion are source-dependent, the extension of this algebra with the priority and unless operators is conservative; see Theorem B.5.1. $\square$

In general, rooted branching bisimulation is not a congruence relation with respect to ACP_τ with guarded linear recursion and the priority and unless operators. We give an example.

Example 7.3.1. Let $A \triangleq \{a, b, c\}$, and let the partial order on $A \cup \{\tau\}$ consist of $\{b < c\}$. We have $a(\tau(b+c)+b) \underline{\leftrightarrow}_{rb} a(b+c)$, because the τ in the process term at the left-hand side is truly silent. However, $\Theta(a(\tau(b+c)+b)) \underline{\leftrightarrow}_{rb} a(\tau c + b)$ and $\Theta(a(b+c)) \underline{\leftrightarrow}_{rb} ac$, so these two process terms are not rooted branching bisimilar, because the τ in the first process term is not truly silent.

Exercise 7.3.2. Explain why the second transition rule for the priority operator cannot be brought into RBB cool format.

A solution to the problem with congruence, suggested by Vaandrager [190], is to give τ priority over any atomic action in A. For instance, if in Example 7.3.1 τ is given priority over any atomic action in A, then

$$\Theta(a(\tau(b+c)+b)) \underline{\leftrightarrow}_{rb} a\tau c \underline{\leftrightarrow}_{rb} ac \underline{\leftrightarrow}_{rb} \Theta(a(b+c)).$$

Theorem 7.3.2. *Let τ have priority over any atomic action in A. Then rooted branching bisimulation equivalence is a congruence with respect to ACP_τ with guarded linear recursion and the priority and unless operators.*

The proof of Theorem 7.3.2 is omitted. An alternative solution to the problem with congruence, suggested by Bergstra and Ponse (see [194]), would be to adapt branching bisimulation equivalence to so-called ι-equivalence, in which non-empty sequences of internal computations may be represented by one silent step: for $k \geq 1$, $a\tau^k b$ is ι-equivalent to $a\tau b$, but not to ab.

In the remainder of this section, τ has priority over any action in A. Table 7.3 presents axioms for the priority and unless operators. The variables x, y, and z range over process terms, while v and w range over $A \cup \{\tau\}$.

Table 7.3. Axioms for priority and unless

TH1		$\Theta(v) = v$
TH2		$\Theta(\delta) = \delta$
TH3		$\Theta(x + y) = \Theta(x) \triangleleft y + \Theta(y) \triangleleft x$
TH4		$\Theta(x \cdot y) = \Theta(x) \cdot \Theta(y)$
P1	$v \not< w$	$v \triangleleft w = v$
P2	$v < w$	$v \triangleleft w = \delta$
P3		$v \triangleleft \delta = v$
P4		$\delta \triangleleft v = \delta$
P5		$(x + y) \triangleleft z = (x \triangleleft z) + (y \triangleleft z)$
P6		$(x \cdot y) \triangleleft z = (x \triangleleft z) \cdot y$
P7		$x \triangleleft (y + z) = (x \triangleleft y) \triangleleft z$
P8		$x \triangleleft (y \cdot z) = x \triangleleft y$

Theorem 7.3.3. $\mathcal{E}_{\text{ACP}_\tau} + \text{RDP}, \text{RSP}, \text{CFAR} + \text{TH1-4} + \text{P1-8}$ *is sound for* ACP_τ *with guarded linear recursion and the priority operator, modulo rooted branching bisimulation equivalence.*

Proof. Since rooted branching bisimulation is both an equivalence and a congruence, we only need to check that if $s = t$ is an axiom and σ a closed substitution that maps the variables in s and t to process terms, then $\sigma(s) \underset{rb}{\leftrightarrow} \sigma(t)$. Here, we only provide some intuition for soundness of the axioms in Table 7.3:

- TH1,2 say that the priority operator leaves atomic actions and the deadlock unchanged, because no behaviour is blocked in these terms;
- TH3 says that in a term $\Theta(s + t)$, initial transitions from $\Theta(s)$ are blocked by initial transitions from t with higher priority, and initial transitions from $\Theta(t)$ are blocked by initial transitions from s with higher priority.
- TH4 says that a term $\Theta(s \cdot t)$ first executes s and then t, and that both in s and in t transitions of a high priority block simultaneous transitions of a lower priority;
- P1,2 are the defining equations for the unless operator; they say that $v \triangleleft w$ can only execute action v if v is not smaller than w;

- P3 says that the deadlock cannot block any actions, while P4 says that $\delta \vartriangleleft v$ does not exhibit any behaviour;
- P5 says that $(s + t) \vartriangleleft u$ can choose between initial transitions of s that are not blocked by initial transitions of u (i.e., $s \vartriangleleft u$), and initial transitions of t that are not blocked by initial transitions of u (i.e., $t \vartriangleleft u$);
- P6,8 say that in $s \vartriangleleft t$ only initial transitions of s are blocked by initial transitions of t;
- P7 says that blocking initial transitions of s by initial transitions of t or u is the same as blocking initial transitions of s by initial transitions of t and blocking the remaining initial transitions of s by initial transitions of u.

These intuitions can be made rigorous by means of explicit rooted branching bisimulation relations between the left- and right-hand sides of closed instantiations of the axioms in Table 7.3. □

Theorem 7.3.4. $\mathcal{E}_{\mathrm{ACP}_\tau} + \mathrm{RDP}, \mathrm{RSP}, \mathrm{CFAR} + \mathrm{TH1\text{-}4} + \mathrm{P1\text{-}8}$ *is complete for* ACP_τ *with guarded linear recursion and the priority and unless operators, modulo rooted branching bisimulation equivalence.*

Proof. It is left to the reader to prove that each process term t is provably equal to a process term $\langle X|E \rangle$ with E a guarded linear recursive specification (see Exercise 7.3.5). Then the desired completeness result follows from the fact that if $\langle X_1|E_1 \rangle \underline{\leftrightarrow}_{rb} \langle Y_1|E_2 \rangle$ for guarded linear recursive specifications E_1 and E_2, then $\langle X_1|E_1 \rangle = \langle Y_1|E_2 \rangle$ can be derived from $\mathcal{E}_{\mathrm{ACP}} + \mathrm{B1}, 2 + \mathrm{RDP}, \mathrm{RSP}$ (see the proof of Theorem 5.3.2). □

Exercise 7.3.3. Let $b < c$, $a < \tau$, $b < \tau$, and $c < \tau$. Derive the equation $\Theta(a(\tau(b + c) + b)) = \Theta(a(b + c))$ from the axioms.

Exercise 7.3.4. Let $\gamma(a, b) \triangleq c$, $a < c$, and $b < c$. Derive

$$\partial_{\{a,b\}}(\langle X \mid X = aX \rangle \| \langle Y \mid Y = bY \rangle) = \Theta(\langle X \mid X = aX \rangle \| \langle Y \mid Y = bY \rangle)$$

from the axioms.

Exercise 7.3.5. Let E be a guarded linear recursive specification and X a recursion variable in E. Prove that $\Theta(\langle X|E \rangle) = \langle Y|F \rangle$ for some guarded linear recursive specification F.

Exercise 7.3.6. Give alternative axioms for the priority operator (without the help of any auxiliary operators), to obtain a sound and complete axiomatization for ACP with linear recursion and the priority operator.

Exercise 7.3.7. The binary *alt* operator in Exercise 3.4.11 alternately executes an action from its first and second argument. Assuming that you answered Exercise 3.4.11, prove that your axioms for the *alt* operator together with $\mathcal{E}_{\mathrm{ACP}_\tau} + \mathrm{RDP}, \mathrm{RSP}, \mathrm{CFAR}$ are complete for ACP_τ with guarded linear recursion and the *alt* operator, modulo rooted branching bisimulation equivalence.

A. Equational Logic

This appendix presents the basic notions of algebraic specification, in which the meaning of terms over some signature is captured using equations. A thorough introduction to the principles of equational logic can be found in [66]. For an overview of algebraic specification of data, see [38, 143].

A.1 Signatures

We start by defining the syntax, which consists of the terms over some algebraic signature.

Definition A.1.1 (Signature). *A signature Σ consists of a finite set of function symbols (or operators) $f, g, \ldots$, where each function symbol f has an arity $ar(f)$, being its number of arguments.*

A function symbol $a, b, c, \ldots$ of arity zero is called a constant, *a function symbol of arity one is called* unary, *and a function symbol of arity two is called* binary.

We assume the presence of a countably infinite set of variables $x, y, z, \ldots$, disjoint from the signature.

Definition A.1.2 (Term). *Let Σ be a signature. The set $\mathbb{T}(\Sigma)$ of (open) terms $s, t, u, \ldots$ over Σ is defined as the least set satisfying:*

- *each variable is in $\mathbb{T}(\Sigma)$;*
- *if $f \in \Sigma$ and $t_1, \ldots, t_{ar(f)} \in \mathbb{T}(\Sigma)$, then $f(t_1, \ldots, t_{ar(f)}) \in \mathbb{T}(\Sigma)$.*

A term is closed *if it does not contain variables. The set of closed terms is denoted by $\mathcal{T}(\Sigma)$.*

We use $o_1 \equiv o_2$ denotes that the objects o_1 and o_1 are syntactically the same. For notational convenience, terms $a()$ are abbreviated to a. For unary function symbols f and natural numbers k, f^k denotes k applications of f: $f^0(t) \equiv t$ and $f^{k+1}(t) \equiv f(f^k(t))$. Often, binary function symbols are represented using infix notation. For example, addition $+$ is a binary function symbol that gives rise to terms $s + t$ (denoting $+(s, t)$).

Exercise A.1.1. Assume a binary function symbol f, a unary function symbol g, and a constant a. Describe the set of closed terms over this signature.

Definition A.1.3 (Substitution). *Let Σ be a signature. A substitution is a mapping σ from variables to the set $\mathbb{T}(\Sigma)$ of open terms. A substitution extends to a mapping from open terms to open terms: the term $\sigma(t)$ is obtained by replacing occurrences of variables x in t by $\sigma(x)$. A substitution σ is closed if $\sigma(x) \in \mathcal{T}(\Sigma)$ for all variables x.*

Exercise A.1.2. Let f be a binary function symbol and a and b constants. Say for the following cases whether there exists a substitution σ such that:

- $\sigma(f(x, y)) \equiv f(a, b)$;
- $\sigma(f(x, x)) \equiv f(a, b)$;
- $\sigma(f(x, y)) \equiv f(z, z)$ and $\sigma(z) \equiv b$;
- $\sigma(g(x)) \equiv \sigma(x)$.

A.2 Axiomatisations

Definition A.2.1 (Axiomatisation). *An* axiomatisation *over a signature Σ is a finite set of equations, called* axioms, *of the form $s = t$ with $s, t \in \mathbb{T}(\Sigma)$.*

An axiomatisation gives rise to an equality relation $=$ on $\mathbb{T}(\Sigma)$.

Definition A.2.2 (Equality relation). *An axiomatisation over a signature Σ induces a binary equality relation $=$ on $\mathbb{T}(\Sigma)$ as follows.*

- (SUBSTITUTION) *If $s = t$ is an axiom and σ a substitution, then $\sigma(s) = \sigma(t)$.*
- (EQUIVALENCE) *The relation $=$ is closed under reflexivity, symmetry, and transitivity:*
  - *$t = t$ for all terms t;*
  - *if $s = t$, then $t = s$;*
  - *if $s = t$ and $t = u$, then $s = u$.*
- (CONTEXT) *The relation $=$ is closed under contexts: if $t = u$ and f is a function symbol with $ar(f) > 0$, then*

$$f(s_1, \ldots, s_{i-1}, t, s_{i+1}, \ldots, s_{ar(f)}) = f(s_1, \ldots, s_{i-1}, u, s_{i+1}, \ldots, s_{ar(f)}).$$

Exercise A.2.1. Let a, b, and c be constants, and f a function symbol of arity three. Consider the axiomatisation

$$f(x, y, z) = f(z, x, y)$$
$$f(x, y, z) = f(y, x, z)$$
$$f(x, c, y) = x$$

Derive the following three equations from the axiomatisation above:

- $f(b, c, a) = f(b, c, b)$;
- $f(a, c, b) = b$;
- $f(c, c, f(c, c, b)) = b$.

A.3 Initial Models

Definition A.3.1 (Model). *Assume an axiomatisation $\mathcal{E}$ over a signature Σ, which induces an equality relation $=$. A model for $\mathcal{E}$ consists of a set $\mathcal{M}$ together with a mapping $\phi : \mathcal{T}(\Sigma) \to \mathcal{M}$.*

- $(\mathcal{M}, \phi)$ *is* sound *for $\mathcal{E}$ if $s = t$ implies $\phi(s) \equiv \phi(t)$ for $s, t \in \mathcal{T}(\Sigma)$;*
- $(\mathcal{M}, \phi)$ *is* complete *for $\mathcal{E}$ if $\phi(s) \equiv \phi(t)$ implies $s = t$ for $s, t \in \mathcal{T}(\Sigma)$.*

Intuitively, the mapping ϕ establishes the interpretation of each closed term in the set $\mathcal{M}$.

Exercise A.3.1. Let the signature consist of a constant a and a unary function symbol f. Say for the following four models whether they are sound and/or complete for the axiomatisation $\{f(x) = f(f(x))\}$:

- $\mathcal{M} \triangleq \{0\}$, and $\phi(f^k(a)) \triangleq 0$ for $k \in \mathbb{N}$;
- $\mathcal{M} \triangleq \{0, 1\}$, $\phi(f^{2k}(a)) \triangleq 0$, and $\phi(f^{2k+1}(a)) \triangleq 1$ for $k \in \mathbb{N}$;
- $\mathcal{M} \triangleq \{0, 1\}$, $\phi(a) \triangleq 0$, and $\phi(f^{k+1}(a)) \triangleq 1$ for $k \in \mathbb{N}$;
- $\mathcal{M} \triangleq \mathbb{N}$, and $\phi(f^k(a)) \triangleq k$ for $k \in \mathbb{N}$.

By the second clause in Definition A.2.2, the equality relation induced by an axiomatisation is by default an equivalence relation. Therefore, it divides the set of closed terms $\mathcal{T}(\Sigma)$ into equivalence classes, where closed terms s and t are in the same equivalence class if and only if $s = t$. The expression $[t]$ denotes the equivalence class that contains the closed term t; that is, $[s]$ and $[t]$ denote the same equivalence class if and only if $s = t$. The set $\{[t] \mid t \in \mathcal{T}(\Sigma)\}$ together with the mapping $\phi(t) \triangleq [t]$ for $t \in \mathcal{T}(\Sigma)$ is a sound and complete model for the axiomatisation, called its *initial model*.

Example A.3.1. As a standard example, we specify the natural numbers with addition and multiplication. The signature consists of the constant 0, the unary successor function S, and the binary functions addition $+$ and multiplication $\cdot$. The equality relation on terms is specified by four axioms:

1. $\quad x + 0 = x$
2. $x + S(y) = S(x + y)$
3. $\quad x \cdot 0 = 0$
4. $x \cdot S(y) = (x \cdot y) + x$

The initial model for this axiomatisation consists of the distinct equivalence classes $[0], [S(0)], [S^2(0)], [S^3(0)], \ldots$. The first three equivalence classes, with some typical representatives of each of these classes, are depicted in Fig. A.1.

Exercise A.3.2. Derive the equation $S(S(S(0))) + S(0) = S(S(0)) \cdot S(S(0))$ (i.e., $3 + 1 = 2 \cdot 2$) from the axiomatisation of the natural numbers.

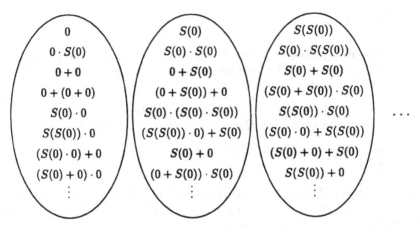

Fig. A.1. Initial model for the natural numbers

Exercise A.3.3. Let a and b be constants and f a unary function symbol. Give the initial models for the following five axiomatisations and signatures:

- $\{x = f(x)\}$ over $\{a, b, f\}$;
- $\{x = f(x)\}$ over $\{a, f\}$;
- the empty axiomatisation $\emptyset$ over $\{a, f\}$;
- $\{x = f(f(x))\}$ over $\{a, f\}$;
- $\{x = f(f(x))\}$ over $\{f\}$.

Assume an axiomatisation over some signature. The function symbols in the signature are well-defined on the equivalence classes in the initial model for the axiomatisation. Namely, if $s_i = t_i$ for $i \in \{1, \ldots, ar(f)\}$, then closure of the equality relation under transitivity and contexts ensures that $f(s_1, \ldots, s_{ar(f)}) = f(t_1, \ldots, t_{ar(f)})$. So the equivalence class $[f(t_1, \ldots, t_{ar(f)})]$ is uniquely determined by the equivalence classes $[t_1]$, $\ldots, [t_{ar(f)}]$. Hence, we can define

$$f([t_1], \ldots, [t_{ar(f)}]) \triangleq [f(t_1, \ldots, t_{ar(f)})].$$

Exercise A.3.4. Show that $S([0]) + S([0])$ and $[S(S(0))]$ represent the same object in the initial model for the axiomatisation over the natural numbers.

Saying that an axiomatisation is ω-complete means that an equation between open terms can be derived from the axiomatisation if all its closed instantiations can be derived from the axiomatisation.

Definition A.3.2 (Omega-completeness). *An axiomatisation $\mathcal{E}$ over a signature Σ is ω-complete if an equation $s = t$ with $s, t \in \mathbb{T}(\Sigma)$ can be derived from $\mathcal{E}$ if $\sigma(s) = \sigma(t)$ can be derived from $\mathcal{E}$ for all closed substitutions σ.*

We note that the axiomatisation of the natural numbers in Example A.3.1 is not ω-complete. For instance, the equation $x + y = y + x$ holds under all closed substitution instances, but this equation cannot be derived from the four axioms in Example A.3.1. Namely, none of these axioms apply to $x + y$ or $y + x$.

Exercise A.3.5. Say for each of the five axiomatisations in Exercise A.3.3 whether it is ω-complete. For each axiomatisation that is not ω-complete, present an equation between open terms that cannot be derived from the axiomatisation, while all its closed instantiations can be derived from the axiomatisation.

A.4 Term Rewriting

A term rewriting system consists of rewrite rules $s \to t$ with s and t open terms, where s is not a single variable and t does not contain fresh variables. Intuitively, a rewrite rule is a directed equation $s = t$ that can only be applied from left to right. An up-to-date overview of term rewriting is given in [8].

Definition A.4.1 (Term rewriting system). *Assume a signature Σ. A rewrite rule is an expression $s \to t$ with $s, t \in \mathbb{T}(\Sigma)$, where:*

1. *the left-hand side s is not a single variable;*
2. *all variables that occur at the right-hand side t also occur in the left-hand side s.*

A term rewriting system (TRS) is a finite set of rewrite rules.

A TRS induces a binary rewrite relation $\to^*$ on terms, similar to the way that an axiomatisation induces an equality relation on terms; see Definition A.2.2. The only distinction is that the rewrite relation is not closed under symmetry, because rewrite rules are directed from left to right.

Definition A.4.2 (Rewrite relation). *A TRS over a signature Σ induces a one-step rewrite relation $\to$ on $\mathbb{T}(\Sigma)$ as follows.*

- *(SUBSTITUTION) If $s \to t$ is a rewrite rule and σ a substitution, then $\sigma(s) \to \sigma(t)$.*
- *(CONTEXT) The relation $\to$ is closed under contexts: if $t \to u$ and f is a function symbol with $ar(f) > 0$, then*

$$f(s_1, \ldots, s_{i-1}, t, s_{i+1}, \ldots, s_{ar(f)}) \to f(s_1, \ldots, s_{i-1}, u, s_{i+1}, \ldots, s_{ar(f)}).$$

The rewrite relation $\to^*$ is the reflexive transitive closure of the one-step rewrite relation $\to$:

- *if $s \to t$, then $s \to^* t$;*

- $t \to^* t$;
- if $s \to^* t$ and $t \to^* u$, then $s \to^* u$.

Example A.4.1. As an example of a TRS, we direct the four equations for natural numbers (see Example A.3.1) from left to right:

1. $\quad x + 0 \to x$
2. $x + S(y) \to S(x + y)$
3. $\quad x \cdot 0 \to 0$
4. $\quad x \cdot S(y) \to (x \cdot y) + x$

Using this TRS, we can prove for instance that $S(0) \cdot S(S(0)) = S(S(0))$, by the following sequence of rewrite steps. In each rewrite step, the subterm that is reduced is underlined.

$$\underline{S(0) \cdot S(S(0))} \overset{(4)}{\to} (\underline{S(0) \cdot S(0)}) + S(0)$$
$$\overset{(4)}{\to} ((\underline{S(0) \cdot 0}) + S(0)) + S(0)$$
$$\overset{(3)}{\to} (\underline{0 + S(0)}) + S(0)$$
$$\overset{(2)}{\to} S(\underline{0 + 0}) + S(0)$$
$$\overset{(1)}{\to} \underline{S(0) + S(0)}$$
$$\overset{(2)}{\to} S(\underline{S(0) + 0})$$
$$\overset{(1)}{\to} S(S(0)).$$

Exercise A.4.1. Derive the equation $S(0) + S(0) = S(0) \cdot S(S(0))$ from the axiomatisation in Example A.3.1.

Term rewriting can be applied to try and compute whether two terms can be equated by an axiomatisation. First, we give a direction to each of the axioms, so that they constitute a TRS. Next, we can try to find a derivation for an equation $s = t$ as follows. Suppose s and t reduce to the same term u: $s \to s_1 \to \cdots \to s_k \to u$ and $t \to t_1 \to \cdots \to t_\ell \to u$. This yields a derivation of $s = t$, owing to the fact that the rewrite rules are directed versions of the axioms: $s = s_1 = \cdots = s_k = u = t_\ell = \cdots = t_1 = t$

Ideally, each reduction of a term by means of a TRS eventually leads to a normal form, which cannot be reduced any further.

Definition A.4.3 (Normal form). *A term is called a* normal form *for a TRS if it cannot be reduced by any of the rewrite rules.*

Definition A.4.4 (Termination). *A TRS is* terminating *if it does not induce infinite reductions* $t_0 \to t_1 \to t_2 \to \cdots$.

Note that the two restrictions on rewrite rules as formulated in Definition A.4.1, the left-hand side is not a single variable and the right-hand side does not contain fresh variables, are essential for termination. Preferably a rewrite relation reduces each term to a unique normal form; that is, if $s \to t_1$ and $s \to t_2$, then both t_1 and t_2 have the same normal form.

Definition A.4.5 (Weak confluence). *A TRS is* weakly confluent *if for each pair of one-step reductions $s \to t_1$ and $s \to t_2$ there is a term u such that $t_1 \to^* u$ and $t_2 \to^* u$.*

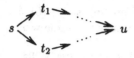

The next lemma from Newman [161] states that termination and weak confluence together are sufficient to guarantee unique normal forms.

Theorem A.4.1 (Newman's lemma). *If a TRS is terminating and weakly confluent, then it reduces each term to a unique normal form.*

Assuming an axiomatisation, we explained previously that one can try to derive an equation $s = t$ by giving a direction to each of the axioms, to obtain a TRS, and attempting to reduce s and t to the same term. If the resulting TRS is terminating and weakly confluent, then this procedure to try and equate s and t is guaranteed to return a derivation if $s = t$. Namely, $s = t$ means that there exists a derivation $s \equiv t_1 = t_2 = \cdots = t_k \equiv t$ in which each equation is the result of an application of an axiom inside a context. Then either $t_i \to t_{i+1}$ or $t_{i+1} \to t_i$ for $i \in \{1, \ldots, k-1\}$. Since the TRS is terminating and weakly confluent, Newman's lemma implies that t_i and t_{i+1} reduce to the same unique normal form for $i \in \{1, \ldots, k-1\}$. So $s \equiv t_1$ and $t \equiv t_k$ reduce to the same unique normal form.

Example A.4.2. The TRS for the natural numbers in Example A.4.1 is terminating. In order to prove this fact, we give an inductive definition of a weight function that maps each term to a natural number.

$$weight(x) \triangleq 1$$
$$weight(0) \triangleq 2$$
$$weight(S(t)) \triangleq weight(t) + 1$$
$$weight(s + t) \triangleq weight(s) + weight(t)^2$$
$$weight(s \cdot t) \triangleq weight(s)^2 \cdot weight(t)^2.$$

It is left to the reader to verify that if $s \to t$ then $weight(s) > weight(t)$. Since each sequence of strictly decreasing natural numbers is finite, it follows that the TRS is terminating.

It is not hard to see that closed terms $s + t$ and $s \cdot t$ are never normal forms, so closed normal forms are of the form $S^k(0)$ for $k \in \mathbb{N}$.

Exercise A.4.2. Prove for the TRS for the natural numbers in Example A.4.1 and for the weight function in Example A.4.2 that if $s \to t$ then $weight(s) > weight(t)$.

Exercise A.4.3. Prove that closed terms of the form $s + t$ or $s \cdot t$ are not normal forms for the TRS for the natural numbers in Example A.4.1.

Exercise A.4.4. Suppose the definition of the weight function in Example A.4.2 would be adapted by putting $weight(0) \triangleq 1$. Give closed terms s and t such that $s \to t$ but s and t have the same weight.

Rewriting Modulo AC. Many axiomatisations from the literature give rise to non-terminating TRSs, which is often due to the fact that they include commutativity and associativity axioms.

Definition A.4.6 (Commutativity and associativity). *Assume an axiomatisation $\mathcal{E}$. A binary function symbol f is* commutative *if $\mathcal{E}$ contains an axiom*

$$f(x, y) = f(y, x)$$

and associative *if $\mathcal{E}$ contains an axiom*

$$f(f(x, y), z) = f(x, f(y, z)).$$

If the equations for commutativity and associativity of a binary function symbol f are turned into rewrite rules, then the resulting TRS is not terminating. For example, if a and b are constants, then the directed version of the commutativity axiom induces the infinite reduction

$$f(a, b) \to f(b, a) \to f(a, b) \to \cdots.$$

This complication can be resolved by applying term rewriting modulo equations (see [170]). That is, we use the equations for commutativity and associativity of f to obtain an equivalence relation $=_{AC}$ on terms: two terms are *equivalent modulo AC of f* if and only if they can be equated by the associativity and commutativity axioms for f. When turning the axiomatisation into a TRS, by giving a direction to the axioms, we exclude the equations for commutativity and associativity of f. Finally, the desired rewrite relation modulo AC of f is obtained by considering terms modulo $=_{AC}$, so that each term actually represents an equivalence class of terms modulo AC of f. This means that $s \to t$ if the TRS induces a one-step reduction $s' \to t'$ where $s =_{AC} s'$ and $t =_{AC} t'$.

Note that it would not be sufficient to work only modulo commutativity of f, because associativity would still give rise to infinite reductions such as:

$$f(f(a, b), c) \to f(a, f(b, c)) =_C f(f(b, c), a) \to f(b, f(c, a)) =_C \cdots.$$

Exercise A.4.5. Let the TRS for the natural numbers in Example A.4.1 be applied to terms modulo AC of the $+$. Reduce the term $S(0) + S(S(0))$ to its normal form in two rewrite steps.

Knuth-Bendix Completion. Axiomatisations can give rise to TRSs that are not weakly confluent. It can be attempted to remedy this imperfection by applying *Knuth-Bendix completion* [135], which determines overlaps in left-hand sides of rewrite rules, and introduces extra rewrite rules to join the resulting right-hand sides (the so-called *critical pairs*). A pair of terms s and t is said to be *convergent* if there exists a term u such that $s \to^* u$ and $t \to^* u$. Knuth-Bendix completion means searching for non-convergent critical pairs, and adding extra rewrite rules in order to make such critical pairs convergent.

Example A.4.3. Let a and b be constants, and f a unary function symbol. Consider the TRS that consists of the following two rewrite rules:

$$a \to b$$
$$f(a) \to b$$

The first rewrite rule induces $f(a) \to f(b)$, while the second rewrite rule induces $f(a) \to b$. The critical pair $f(b)$ and b is not convergent. This pair can be made convergent by adding an extra rewrite rule to the TRS:

$$f(b) \to b$$

(Note that the reverse rewrite rule, $b \to f(b)$, would produce a non-terminating TRS.) The resulting TRS is weakly confluent and terminating.

Exercise A.4.6. Let a be a constant, g and h unary function symbols, and f a binary function symbol. Consider the TRS that consists of the following two rewrite rules:

$$g(f(h(x), x)) \to h(x)$$
$$f(x, a) \to x$$

Determine the non-convergent critical pairs, and apply Knuth-Bendix completion to obtain a TRS that is weakly confluent and terminating.

See [94] for an application of Knuth-Bendix completion in the realm of process algebra with iteration operators. The significance of making critical pairs convergent is expressed by the following theorem, due to Huet [130].

Theorem A.4.2. *A TRS is weakly confluent if and only if all its critical pairs are convergent.*

B. Structural Operational Semantics

This appendix introduces the basics of structural operational semantics [171], which defines a labelled transition system over a term algebra. An up-to-date overview of structural operational semantics is given in [3].

B.1 Transition System Specifications

We assume a non-empty set S of states, together with a finite, non-empty set of transition labels A and a finite set of predicate symbols.

Definition B.1.1 (Labelled transition system). *A transition is a triple (s, a, s') with $a \in A$, or a pair (s, P) with P a predicate, where $s, s' \in S$. A labelled transition system (LTS) is a (possibly infinite) set of transitions. An LTS is finitely branching if each of its states has only finitely many outgoing transitions.*

For convenience of notation, a transition (s, a, s') is usually denoted as $s \xrightarrow{a} s'$; it expresses that the state s can evolve into the state s' by the execution of action a. Moreover, a transition (s, P) is usually denoted as sP; it expresses that predicate P holds in the state s.

In this text, the states of an LTS are always the closed terms (see Definition A.1.2) over a signature Σ (see Definition A.1.1). In other words, transitions are expressions $t \xrightarrow{a} t'$ and tP with $t, t' \in \mathcal{T}(\Sigma)$. In view of the syntactic structure of closed terms over a signature, such transitions can be derived by means of inductive proof rules, where the validity of a number of transitions (the premises) may imply the validity of some other transition (the conclusion).

Definition B.1.2 (Transition system specification). *A transition rule ρ is an expression of the form $\frac{H}{\pi}$, with H a set of expressions $t \xrightarrow{a} t'$ and tP with $t, t' \in \mathbb{T}(\Sigma)$, called the (positive) premises of ρ, and π an expression $t \xrightarrow{a} t'$ or tP with $t, t' \in \mathbb{T}(\Sigma)$, called the conclusion of ρ. The left-hand side of π is called the source of ρ. A transition rule is closed if it does not contain any variables.*

A transition system specification (TSS) is a (possibly infinite) set of transition rules.

We want to give meaning to TSSs; that is, each TSS is to generate an LTS. For this purpose we use the notion of a *proof* of a closed transition rule from a TSS.

Definition B.1.3 (Proof). *A proof from a TSS T of a closed transition rule $\frac{H}{\pi}$ consists of an upwardly branching tree in which all upward paths are finite, where the nodes of the tree are labelled by transitions such that:*

- *the root has label π;*
- *if some node has label ℓ, and K is the set of labels of nodes directly above this node, then*
  1. *either K is the empty set and $\ell \in H$,*
  2. *or $\frac{K}{\ell}$ is a closed substitution instance of a transition rule in T.*

Definition B.1.4 (Generated LTS). *We define that the LTS generated by a TSS T consists of the transitions π such that $\frac{\emptyset}{\pi}$ can be proved from T.*

For notational convenience, the premises of a transition rule are not always presented using proper set notation.

Example B.1.1. Let the signature consist of a constant a and a unary function symbol f. The TSS

$$\frac{}{aP} \qquad \frac{xP}{f(x)P}$$

generates the LTS $\{f^k(a)P \mid k \in \mathbb{N}\}$. The proof of $\frac{}{f^k(a)P}$ for $k \in \mathbb{N}$ is

$$
\begin{array}{c}
aP \bullet \\
\downarrow \\
f(a)P \bullet \\
\downarrow \\
\vdots \\
\downarrow \\
f^k(a)P \bullet
\end{array}
$$

Exercise B.1.1. Let the signature consist of constants a and b and a unary function symbol f. Give the LTSs that are generated by the following TSSs:

- $\dfrac{xP}{f(x)P}$

- $\dfrac{}{aP} \qquad \dfrac{xP}{f(x)P} \qquad \dfrac{bQ}{bQ}$

- $\dfrac{aP}{bQ} \qquad \dfrac{bQ}{aP}$

- $\dfrac{}{aP} \qquad \dfrac{aP}{bQ} \qquad \dfrac{bQ}{aP}$

B.2 The Meaning of Negative Premises

Sometimes it is useful to allow *negative* premises of the form $t \overset{a}{\nrightarrow}$ or $t \neg P$ in transition rules. Intuitively, a closed substitution instance σ of such a negative premise is valid if $\sigma(t) \overset{a}{\to} t'$ does not hold for any closed term t', or if $\sigma(t)P$ does not hold, respectively.

It is not always clear which LTS is generated by a TSS that contains transition rules with negative premises. For example, the transition rule

$$\frac{a \neg P}{aP}$$

expresses that aP holds if aP does not hold. On the one hand this excludes the possibility that aP does not hold, but on the other hand it does not establish a firm proof for aP. Therefore, on the basis of the transition rule above it is unknown whether the transition aP holds.

Three-Valued Stable Models. The *three-valued stable model*, introduced by Baeten, Bergstra, Klop, and Weijland [23] in term rewriting and by Przymusinski [175] in logic programming, can be used to give meaning to TSSs with negative premises. A three-valued stable model partitions the collection of transitions into three disjoint sets: the set C of transitions that are true, the set F of transitions that are false, and the set U of transitions for which it is unknown whether or not they are true. Such a partitioning is determined by the pair $\langle C, U \rangle$.

We want to extend Definition B.1.3 for a proof of a closed transition rule from a TSS to the setting with negative premises. Therefore, from now on we allow the proof tree in Definition B.1.3 to contain expressions $t \overset{a}{\nrightarrow}$ and $t \neg P$ as labels of its nodes, where t is a closed term.

Definition B.2.1. *A set N of expressions $t \overset{a}{\nrightarrow}$ and $t \neg P$ (where t ranges over closed terms, a over A, and P over predicates) holds for a set S of transitions, denoted by $S \models N$, if:*

1. *for each $t \overset{a}{\nrightarrow} \in N$ we have that $t \overset{a}{\to} t' \notin S$ for all $t' \in \mathcal{T}(\Sigma)$;*
2. *for each $t \neg P \in N$ we have that $tP \notin S$.*

Definition B.2.2 (Three-valued stable model). *A pair $\langle C, U \rangle$ of disjoint sets of transitions is a* three-valued stable model *for a TSS T if it satisfies the following two requirements:*

1. *a transition π is in C if and only if T proves a closed transition rule $\frac{N}{\pi}$ where N contains only negative premises and $C \cup U \models N$;*
2. *a transition π is in $C \cup U$ if and only if T proves a closed transition rule $\frac{N}{\pi}$ where N contains only negative premises and $C \models N$.*

Example B.2.1. Let the signature consist of constants a and b. The TSS

$$\frac{a\neg P}{bQ} \qquad \frac{b\neg Q}{aP}$$

has the following three-valued stable models:

- $\langle \emptyset, \{aP, bQ\}\rangle$;
- $\langle \{aP\}, \emptyset\rangle$;
- $\langle \{bQ\}, \emptyset\rangle$.

Exercise B.2.1. Let the signature consist of constants a and b. Give the three-valued stable models for the following TSSs:

- $$\frac{a\neg P}{bQ}$$

- $$\frac{a\neg P}{aP}$$

- $$\frac{}{aP} \qquad \frac{a\neg P}{aP}$$

- $$\frac{aP}{aP} \qquad \frac{a\neg P}{bQ} \qquad \frac{b\neg Q}{aP}$$

- $$\frac{x\neg P \quad a\neg Q}{xQ} \qquad \frac{x\neg Q \quad b\neg P}{xP}$$

Least Three-Valued Stable Models. Each TSS T allows a *least* three-valued stable model $\langle C, U\rangle$, in the sense that the sets C and F are both minimal and the set U is maximal. This least three-valued stable model coincides with the so-called well-founded model [97] (see [175]). The construction of the least three-valued stable model for a TSS uses the notion of ordinal numbers $\alpha, \beta, \gamma, \ldots$, which form an ordered set that extends the natural numbers.

Definition B.2.3 (Ordinal number). *The* ordinal numbers *are defined inductively by:*

1. *0 is the smallest ordinal number;*
2. *each ordinal number α has a successor $\alpha + 1$;*
3. *each sequence of ordinal numbers $\alpha < \alpha + 1 < \alpha + 2 < \cdots$ is capped by a limit ordinal λ.*

A limit ordinal does not have a direct predecessor. The first ordinal numbers are the natural numbers $0 < 1 < 2 < \cdots$, which give rise to the limit ordinal ω. The successors of this limit ordinal give rise to a sequence of ordinal numbers $\omega < \omega + 1 < \omega + 2 < \cdots$, resulting in the limit ordinal 2ω, et cetera. Similar to standard induction on the natural numbers, one can apply *ordinal induction* over the set of ordinal numbers. That is, to prove that a property

P_α holds for all ordinal numbers α, it suffices to prove that if P_β holds for all ordinal numbers $\beta < \gamma$, then P_γ holds.

The least three-valued stable model for a TSS T can be constructed as follows. First we define a sequence $\langle C_\alpha, U_\alpha \rangle$ of pairs of disjoint sets of transitions for ordinal numbers α, using ordinal induction.

- $C_0 = \emptyset$ and U_0 contains all transitions.
- For ordinal numbers α, $\langle C_{\alpha+1}, U_{\alpha+1} \rangle$ is constructed from $\langle C_\alpha, U_\alpha \rangle$ as follows.
  A transition π is in $C_{\alpha+1}$ if and only if T proves a closed transition rule $\frac{N}{\pi}$ where N contains only negative premises and $C_\alpha \cup U_\alpha \models N$.
  A transition π is in $C_{\alpha+1} \cup U_{\alpha+1}$ if and only if T proves a closed transition rule $\frac{N}{\pi}$ where N contains only negative premises and $C_\alpha \models N$.
- For limit ordinals α we define $C_\alpha = \cup_{\beta<\alpha} C_\beta$ and $U_\alpha = \cap_{\beta<\alpha} U_\beta$.

The construction of C_α and U_α for ordinal numbers α is such that if $\alpha \leq \beta$ then $C_\alpha \subseteq C_\beta$ and $U_\alpha \supseteq U_\beta$. So the Knaster-Tarski fixed point theorem [184] ensures that there is an ordinal number γ such that C_γ is maximal and U_γ is minimal; in other words, $C_{\gamma+1}$ and $U_{\gamma+1}$ coincide with C_γ and U_γ, respectively. From this observation, together with the construction of C_γ and U_γ, it follows that $\langle C_\gamma, U_\gamma \rangle$ is a three-valued stable model for T. Furthermore, if $\langle C', U' \rangle$ is some three-valued stable model for T, then it follows by ordinal induction that $U' \subseteq U_\alpha$ for all ordinal numbers α, so in particular $U' \subseteq U_\gamma$. Hence, $\langle C_\gamma, U_\gamma \rangle$ is the least three-valued stable model for T.

Example B.2.2. For the TSS in Example B.2.1,

$$\frac{a \neg P}{bQ} \qquad \frac{b \neg Q}{aP}$$

we have that C_α is $\emptyset$ for $\alpha \geq 0$, U_0 is $\{aP, aQ, bP, bQ\}$, and U_α is $\{aP, bQ\}$ for $\alpha \geq 1$. So its least three-valued stable model is $\langle \emptyset, \{aP, bQ\} \rangle$.

Exercise B.2.2. For each TSS in Exercise B.2.1, construct the sequence of pairs $\langle C_\alpha, U_\alpha \rangle$ for ordinal numbers α, and conclude from these pairs what is the least three-valued stable model.

Positive after Reduction. Bol and Groote [59] introduced the notion of a TSS that is positive after reduction; the definition below stems from [103].

Definition B.2.4 (Positive after reduction). *A TSS is positive after reduction if its least three-valued stable model does not contain unknown transitions.*

If a TSS is positive after reduction, then it allows only one three-valued stable model, which by default is its least one; see [103]. We define that the LTS generated by a TSS T that is positive after reduction consists of the transitions that are true in the three-valued stable mode of T. A TSS that does not contain transition rules with negative premises is always positive after reduction, and the LTS that such a TSS generates according to Definition B.1.4 coincides with the set of true transitions in its three-valued stable model; see [103].

Exercise B.2.3. Say for each of the TSSs in Exercise B.2.1 whether it is positive after reduction.

Stratifications. A useful tool for showing that a TSS is positive after reduction is the notion of a *stratification* [108, 174]. Basically, a TSS is stratified if there exists a weight function on transitions such that for each closed substitution and for each transition rule, the substitution instances of the positive premises are smaller than or equal to the substitution instance of the conclusion, and the substitution instances of the negative premises are strictly smaller than the substitution instance of the conclusion.

Definition B.2.5 (Stratification). *A stratification for a TSS is a weight function ϕ which maps transitions to ordinal numbers, such that for each transition rule ρ with conclusion π and for each closed substitution σ:*

1. *for positive premises $t \xrightarrow{a} t'$ and tP of ρ, $\phi(\sigma(t) \xrightarrow{a} \sigma(t')) \leq \phi(\sigma(\pi))$ and $\phi(\sigma(t)P) \leq \phi(\sigma(\pi))$, respectively;*
2. *for negative premises $t \xarrownot{a}$ and $t\neg P$ of ρ, $\phi(\sigma(t) \xrightarrow{a} t') < \phi(\sigma(\pi))$ for all closed terms t' and $\phi(\sigma(t)P) < \phi(\sigma(\pi))$, respectively.*

The following result stems from [59].

Theorem B.2.1. *If a TSS allows a stratification, then it is positive after reduction.*

Exercise B.2.4. Say for each of the TSSs in Exercise B.2.1 whether it allows a stratification.

Exercise B.2.5. Give a counter-example to show that the reverse of Theorem B.2.1 does not hold: a TSS that is positive after reduction may not always allow a stratification.

Exercise B.2.6. Let the signature consist of a constant a and a unary function symbol f. Moreover, let P and Q be predicates. Give a stratification for the TSS

$$\frac{x\neg P}{xQ} \qquad \frac{x\neg P}{f(x)P}$$

Give its three-valued stable model.

B.3 Bisimulation as a Congruence

A process graph, or process for short, is an LTS that starts its execution (i.e., evolving from state to state) in a designated root state.

Definition B.3.1 (Process graph). *A process (graph) p is an LTS in which one state s is elected to be the* root. *If the LTS contains a transition $s \xrightarrow{a} s'$, then $p \xrightarrow{a} p'$ where p' has root state s'. Moreover, if the LTS contains a transition sP, then pP.*

- *A process p_0 is* finite *if there are only finitely many sequences $p_0 \xrightarrow{a_1} p_1 \xrightarrow{a_2} \cdots \xrightarrow{a_k} p_k$.*
- *A process p_0 is* regular *if there are only finitely many processes p_k such that $p_0 \xrightarrow{a_1} p_1 \xrightarrow{a_2} \cdots \xrightarrow{a_k} p_k$.*

Exercise B.3.1. Give processes that satisfy the following requirements, respectively:

- finite and regular;
- neither finite nor regular;
- regular but not finite.

A wide range of semantic equivalences have been developed to distinguish process graphs. Classic process equivalences are trace equivalence and simulation equivalence. More recently conceived process equivalences include ready equivalence [46, 162], failure equivalence [64, 46], ready trace equivalence [20, 173], failure trace equivalence [169], ready simulation [58, 141], and testing equivalences [80, 120]. See [99] for an overview and comparison of existing process equivalences.

This text focuses on *bisimulation equivalence* [30, 153, 166], which is finer than any of the process equivalences mentioned above; that is, if two processes are equivalent with respect to bisimulation, then they are also equivalent modulo any of the aforementioned equivalences. Bisimulation equivalence not only requires that two processes can execute the same strings of transitions, but also that they have the same branching structure. See Section 2.3 for an exposition on why bisimulation makes a sensible equivalence relation to distinguish process behaviour in a setting with concurrency.

Definition B.3.2 (Bisimulation). *A bisimulation relation $\mathcal{B}$ is a binary relation on processes such that:*

1. *if $p \mathcal{B} q$ and $p \xrightarrow{a} p'$, then $q \xrightarrow{a} q'$ with $p' \mathcal{B} q'$;*
2. *if $p \mathcal{B} q$ and $q \xrightarrow{a} q'$, then $p \xrightarrow{a} p'$ with $p' \mathcal{B} q'$;*
3. *if $p \mathcal{B} q$ and pP, then qP;*
4. *if $p \mathcal{B} q$ and qP, then pP.*

Two processes p and q are bisimilar, *denoted by $p \leftrightarrow q$, if there is a bisimulation relation $\mathcal{B}$ such that $p \mathcal{B} q$.*

It is not hard to check that $\leftrightarrow$ is an equivalence relation :

- $\leftrightarrow$ is reflexive, i.e., $p \leftrightarrow p$;
- $\leftrightarrow$ is symmetric, i.e., if $p \leftrightarrow q$ then $q \leftrightarrow p$;
- $\leftrightarrow$ is transitive, i.e., if $p \leftrightarrow q$ and $q \leftrightarrow r$, then $p \leftrightarrow r$.

Hence, $\leftrightarrow$ divides the collection of processes into equivalence classes.

Example B.3.1. The following two processes, with root states s_0 and s_4, respectively, are bisimilar:

$$\{s_0 \xrightarrow{a} s_1,\ s_1 \xrightarrow{b} s_2,\ s_1 \xrightarrow{b} s_3\}$$
$$\{s_4 \xrightarrow{a} s_5,\ s_4 \xrightarrow{a} s_6,\ s_5 \xrightarrow{b} s_7,\ s_6 \xrightarrow{b} s_7\}.$$

A bisimulation relation between these two processes is defined by: $s_0 \mathcal{B} s_4$, $s_1 \mathcal{B} s_5$, $s_1 \mathcal{B} s_6$, $s_2 \mathcal{B} s_7$, and $s_3 \mathcal{B} s_7$. This bisimulation relation can be depicted as follows:

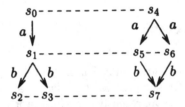

Exercise B.3.2. Give a bisimulation relation between the processes $\{s_0 \xrightarrow{a} s_0,\ s_0 \xrightarrow{b} s_1,\ s_1 \xrightarrow{b} s_0,\ s_1 \xrightarrow{a} s_1\}$ and $\{s \xrightarrow{a} s,\ s \xrightarrow{b} s\}$, with root states s_0 and s, respectively. Prove that this relation is indeed a bisimulation relation.

Kanellakis and Smolka [132] presented an efficient algorithm to decide whether two regular processes (see Definition B.3.1) are bisimilar. Paige and Tarjan [164] presented an alternative algorithm with an even better worst-case time complexity. In contrast with other process equivalences, bisimulation equivalence is decidable for normed processes, in which from each state there is a finite sequence of transitions that leads to successful termination; see [22, 67, 123, 124].

In the remainder of this section we assume a term algebra over some signature.

Definition B.3.3 (Congruence). *Let Σ be a signature. An equivalence relation $\mathcal{B}$ on $\mathcal{T}(\Sigma)$ is a congruence if for each $f \in \Sigma$,*

if $s_i \mathcal{B} t_i$ for $i \in \{1, \dots, ar(f)\}$, then $f(s_1, \dots, s_{ar(f)}) \mathcal{B} f(t_1, \dots, t_{ar(f)})$.

Congruence is an essential property for bisimulation equivalence over a term algebra, in order to be able to give an axiomatisation that is sound and complete modulo bisimulation; i.e., $s = t$ if and only if $s \leftrightarrow t$ (cf. Definition

A.3.1). Namely, the equality relation over a term algebra as induced by an axiomatisation is by default closed under contexts; see Definition A.2.2.

Bisimulation equivalence with respect to the LTS generated by a TSS is not necessarily a congruence. Groote and Vaandrager [116] introduced a syntactic format for TSSs, which was extended with negative premises [108] and predicates [26, 192] to obtain the so-called *panth* format. If a TSS is positive after reduction and in panth format, then the bisimulation equivalence that it induces is guaranteed to be a congruence.

Definition B.3.4 (Panth format). *A transition rule ρ is in* panth *format if it satisfies the following three restrictions:*

1. *for each positive premise $t \xrightarrow{a} t'$ of ρ, the right-hand side t' is a single variable;*
2. *the source of ρ contains no more than one function symbol; and*
3. *there are no multiple occurrences of the same variable at the right-hand sides of positive premises and in the source of ρ.*

A TSS is said to be in panth *format if it consists of panth rules only.*

Theorem B.3.1. *If a TSS is positive after reduction and in panth format, then the bisimulation equivalence that it induces is a congruence.*

Proof. See [59, 91].

We give an example to show that the restriction in Theorem B.3.1 to TSSs that are positive after reduction is essential. In particular, it cannot be relaxed to TSSs that have exactly one (not necessarily least) three-valued stable model that does not contain unknown transitions. The example is derived from Example 8.12 in [59].

Example B.3.2. Let the signature consist of constants a and b and a unary function symbol f. Consider the following TSS in panth format:

$$\frac{}{aP} \quad \frac{}{bP} \quad \frac{xP \quad f(x)\neg Q_1 \quad f(a)\neg Q_2}{f(x)Q_2} \quad \frac{xP \quad f(x)\neg Q_2 \quad f(b)\neg Q_1}{f(x)Q_1}$$

Its least three-valued stable model contains as true transitions aP and bP, and as unknown transitions $f(a)Q_1$, $f(a)Q_2$, $f(b)Q_1$, and $f(b)Q_2$ (cf. the fourth TSS in Exercise B.2.1). So the TSS is not positive after reduction.

The TSS has a second three-valued stable model, in which $aP, bP, f(a)Q_1$, and $f(b)Q_2$ are the true transitions, and the set of unknown transitions is empty. We have $a \leftrightarrow b$ and $f(a) \not\leftrightarrow f(b)$ with respect to the latter three-valued stable model. So the induced bisimulation equivalence is not a congruence.

Exercise B.3.3. Verify that the TSS in Exercise B.2.6 is in panth format. Show that the bisimulation equivalence induced by this TSS is a congruence.

Exercise B.3.4. Let the signature consist of constants a and b, a unary function symbol f, and a binary function symbol g. Show that bisimulation equivalence as induced by each of the TSSs below is not a congruence. Which syntactic requirements of the panth format do these respective TSSs violate?

- $$\frac{}{f(a)P}$$

- $$\frac{x\neg P}{g(x,x)P}$$

- $$\frac{}{x \xrightarrow{c} a} \qquad \frac{x \xrightarrow{c} y}{f(y)P}$$

- $$\frac{}{x \xrightarrow{c} x} \qquad \frac{x \xrightarrow{c} a}{f(x)P}$$

- $$\frac{}{x \xrightarrow{c} x} \qquad \frac{x_1 \xrightarrow{c} y \quad x_2 \xrightarrow{c} y}{g(x_1,x_2)P}$$

B.4 Branching Bisimulation as a Congruence

In order to abstract away from internal actions, Milner [151] introduced a special transition label τ, called the *silent step*. A number of equivalence notions have been developed to distinguish processes that incorporate silent steps, such as delay equivalence [152], η-equivalence [25], and observation equivalence [155]. See [100] for an overview and comparison of process equivalences in the presence of the silent step.

This text focuses on *branching bisimulation equivalence* [105], which is finer than any of the process equivalences mentioned above. Intuitively, branching bisimulation equivalence allows us to abstract away from a τ-transition if its execution does not lose any possible behaviour. See Section 5.1 and [102] for expositions on why rooted branching bisimulation makes a sensible equivalence relation to abstract away from internal computations.

Definition B.4.1 (Branching bisimulation). *A branching bisimulation relation B is a binary relation on the collection of processes such that:*

1. *if $p B q$ and $p \xrightarrow{a} p'$, then*
   - *either $a \equiv \tau$ and $p' B q$;*
   - *or there is a sequence of (zero or more) τ-transitions $q \xrightarrow{\tau} \cdots \xrightarrow{\tau} q_0$ such that $p B q_0$ and $q_0 \xrightarrow{a} q'$ with $p' B q'$;*
2. *if $p B q$ and $q \xrightarrow{a} q'$, then*
   - *either $a \equiv \tau$ and $p B q'$;*
   - *or there is a sequence of (zero or more) τ-transitions $p \xrightarrow{\tau} \cdots \xrightarrow{\tau} p_0$ such that $p_0 B q$ and $p_0 \xrightarrow{a} p'$ with $p' B q'$.*

3. if pBq and pP, then there is a sequence of (zero or more) τ-transitions $q \xrightarrow{\tau} \cdots \xrightarrow{\tau} q_0$ such that pBq_0 and q_0P;

4. if pBq and qP, then there is a sequence of (zero or more) τ-transitions $p \xrightarrow{\tau} \cdots \xrightarrow{\tau} p_0$ such that p_0Bq and p_0P.

Two processes p and q are branching bisimilar, *denoted by $p \leftrightarrow_b q$, if there is a branching bisimulation relation B such that pBq.*

The relation $\leftrightarrow_b$ is an equivalence on the collection of processes; see [32]. Groote and Vaandrager [115] presented an efficient algorithm to decide whether two regular processes are branching bisimilar.

Example B.4.1. The following two processes, with root states s_0 and s_2, respectively, are branching bisimilar:

$$\{s_0 \xrightarrow{\tau} s_0, \; s_0 \xrightarrow{a} s_1, \; s_0 \xrightarrow{b} s_1\}$$
$$\{s_2 \xrightarrow{a} s_3, \; s_2 \xrightarrow{\tau} s_4, \; s_4 \xrightarrow{a} s_5, \; s_4 \xrightarrow{b} s_5\}.$$

A branching bisimulation relation between these two processes is defined by: $s_0 B s_2$, $s_0 B s_4$, $s_1 B s_3$, and $s_1 B s_5$. This branching bisimulation relation can be depicted as follows:

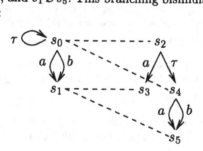

Exercise B.4.1. Say for each of the following four pairs of processes, with root states s_0 and s_2, respectively, whether they are branching bisimilar:

- $\{s_0 \xrightarrow{\tau} s_0, \; s_0P\}$ and $\{s_2P\}$;
- $\{s_0 \xrightarrow{\tau} s_1, \; s_1P, \; s_1Q\}$ and $\{s_2P, \; s_2Q\}$;
- $\{s_0 \xrightarrow{\tau} s_1, \; s_0P, \; s_1Q\}$ and $\{s_2P, \; s_2Q\}$;
- $\{s_0 \xrightarrow{\tau} s_0, \; s_0 \xrightarrow{a} s_1, \; s_1 \xrightarrow{\tau} s_1\}$ and $\{s_2 \xrightarrow{\tau} s_3, \; s_2 \xrightarrow{a} s_4, \; s_3 \xrightarrow{a} s_4\}$.

In each case, either give a branching bisimulation relation or explain why the processes are not branching bisimilar.

A rootedness condition, originating from [155], is imposed on top of branching bisimulation, in order to make it a congruence with respect to a fundamental operation, which joins process graphs at their root states (see Section 5.1).

Definition B.4.2 (Rooted branching bisimulation). *A* rooted branching bisimulation relation B *is a binary relation on processes such that:*

1. *if $p \mathcal{B} q$ and $p \xrightarrow{a} p'$, then $q \xrightarrow{a} q'$ with $p' \underline{\leftrightarrow}_b q'$;*
2. *if $p \mathcal{B} q$ and $q \xrightarrow{a} q'$, then $p \xrightarrow{a} p'$ with $p' \underline{\leftrightarrow}_b q'$;*
3. *if $p \mathcal{B} q$ and pP, then qP;*
4. *if $p \mathcal{B} q$ and qP, then pP.*

Two processes p and q are rooted branching bisimilar, *denoted by $p \underline{\leftrightarrow}_{rb} q$, if there is a* rooted branching bisimulation relation $\mathcal{B}$ *such that $p \mathcal{B} q$.*

Since branching bisimilarity is an equivalence relation, it is not hard to see that rooted branching bisimilarity is also an equivalence relation. Hence, $\underline{\leftrightarrow}_{rb}$ divides the collection of processes into equivalence classes.

Exercise B.4.2. Show that the processes $\{ s_0 \xrightarrow{a} s_1, \ s_1 \xrightarrow{b} s_2, \ s_1 \xrightarrow{\tau} s_3, \ s_3 \xrightarrow{b} s_4, \ s_3 \xrightarrow{c} s_2 \}$ and $\{ s_5 \xrightarrow{a} s_6, \ s_6 \xrightarrow{b} s_7, \ s_6 \xrightarrow{c} s_7 \}$, with root states s_0 and s_5, respectively, are rooted branching bisimilar.

Exercise B.4.3. Say for each of the pairs of processes in Exercise B.4.1 whether they are rooted branching bisimilar.

If a TSS is positive after reduction and in *RBB cool* format [57, 90], then the rooted branching bisimulation equivalence that it induces is guaranteed to be a congruence. We proceed to present the RBB cool format, which requires two auxiliary definitions.

Definition B.4.3 (Lookahead). *A transition rule contains* lookahead *if a variable occurs at the left-hand side of a premise and at the right-hand side of a premise of this rule.*

Definition B.4.4 (Patience rule). *A* patience rule for the i-th argument *of a function symbol f is a panth rule of the form*

$$\frac{x_i \xrightarrow{\tau} y}{f(x_1, \ldots, x_{ar(f)}) \xrightarrow{\tau} f(x_1, \ldots, x_{i-1}, y, x_{i+1}, \ldots, x_{ar(f)})}$$

Definition B.4.5 (RBB cool format). *A TSS T is in RBB cool format if the following requirements are fulfilled.*

1. *T consists of panth rules that do not contain lookahead.*
2. *Suppose a function symbol f occurs at the right-hand side of the conclusion of some transition rule in T. Let $\rho \in T$ be a non-patience rule with source $f(x_1, \ldots, x_{ar(f)})$. Then for $i \in \{1, \ldots, ar(f)\}$, x_i occurs in no more than one premise of ρ, where this premise is of the form $x_i P$ or $x_i \xrightarrow{a} y$ with $a \not\equiv \tau$. Moreover, if there is such a premise in ρ, then there is a patience rule for the i-th argument of f in T.*

Theorem B.4.1. *If a TSS is positive after reduction and in RBB cool format, then the rooted branching bisimulation equivalence that it induces is a congruence.*

Proof. See [57, 90].

The following counter-example and exercise show the need for the restrictions of the RBB cool format on top of the panth format.

Example B.4.2. Let the signature consist of constants a, b, c, and d, together with a unary function symbol f. Consider the following TSS in panth format:

$$\frac{}{a \xrightarrow{a} c} \qquad \frac{}{b \xrightarrow{a} d} \qquad \frac{}{c \xrightarrow{\tau} d} \qquad \frac{}{dP} \qquad \frac{x \xrightarrow{a} y \quad yP}{f(x)P}$$

It is not hard to see that $a \underline{\leftrightarrow}_{rb} b$. However, $f(a) \not\underline{\leftrightarrow}_{rb} f(b)$, because $f(b)P$ holds while $f(a)P$ does not hold. So the rooted branching bisimulation equivalence induced by the TSS above is not a congruence.

The TSS does not contain negative premises, so by default it is positive after reduction. Hence, Theorem B.4.1 implies that the TSS cannot be in RBB cool format. Note that the fifth transition rule violates the RBB cool restriction that the transition rules must not contain lookahead.

Exercise B.4.4. Let the signature consist of constants a, b, c, d, and e, together with a unary function symbol f. Show for each of the following TSSs in panth format that $a \underline{\leftrightarrow}_{rb} b$ but $f(a) \not\underline{\leftrightarrow}_{rb} f(b)$. Which RBB cool restrictions do these respective TSSs violate?

- $$\frac{}{a \xrightarrow{a} c} \quad \frac{}{b \xrightarrow{a} d} \quad \frac{}{c \xrightarrow{\tau} d} \quad \frac{}{dP} \quad \frac{x \xrightarrow{a} y}{f(x) \xrightarrow{a} f(y)} \quad \frac{xP}{f(x)P}$$

- $$\frac{}{a \xrightarrow{\tau} c} \quad \frac{}{b \xrightarrow{\tau} d} \quad \frac{}{c \xrightarrow{\tau} d} \quad \frac{}{dP} \quad \frac{x \xrightarrow{\tau} y}{f(x) \xrightarrow{\tau} f(y)} \quad \frac{x \neg P}{f(x)P}$$

- $$\frac{}{a \xrightarrow{\tau} c} \quad \frac{}{b \xrightarrow{\tau} d} \quad \frac{}{cP} \quad \frac{}{c \xrightarrow{\tau} e} \quad \frac{}{eQ} \quad \frac{}{e \xrightarrow{\tau} c} \quad \frac{}{dP} \quad \frac{}{dQ}$$

$$\frac{x \xrightarrow{\tau} y}{f(x) \xrightarrow{\tau} f(y)} \quad \frac{xP \quad xQ}{f(x)P}$$

Exercise B.4.5. Suppose the first TSS in Exercise B.4.4 is extended with the patience rule $\frac{x \xrightarrow{\tau} y}{f(x) \xrightarrow{\tau} f(y)}$ for the argument of f. Prove that the resulting TSS is in RBB cool format. Verify that $f(a) \underline{\leftrightarrow}_{rb} f(b)$.

B.5 Conservative Extension

Let a TSS T_0 over some signature Σ_0 be extended with a TSS T_1 over some signature Σ_1. We want the resulting TSS $T_0 \oplus T_1$ over the signature $\Sigma_0 \oplus \Sigma_1$ to be a conservative extension of T_0, meaning that the transition rules in T_1

do not affect the transitions of closed terms over Σ_0. In order to be able to combine T_0 and T_1, it is assumed in the remainder of this section that the function symbols in the intersection of their signatures have the same arity in both signatures.

Definition B.5.1 (Conservative extension). *Let T_0 and T_1 be TSSs over signatures Σ_0 and Σ_1, respectively. The TSS $T_0 \oplus T_1$ is a conservative extension of T_0 if the LTSs generated by T_0 and $T_0 \oplus T_1$ contain exactly the same transitions $t \xrightarrow{a} t'$ and tP with $t \in \mathcal{T}(\Sigma_0)$.*

Exercise B.5.1. Let a and b be constants. Say for each of the following pairs of TSSs T_0 and T_1 over the signatures $\{a\}$ and $\{a, b\}$, respectively, whether $T_0 \oplus T_1$ is a conservative extension of T_0. In cases where the extension is not conservative, give a transition of a that holds with respect to $T_0 \oplus T_1$ but not with respect to T_0.

- $\emptyset$ and $\dfrac{}{aP}$

- $\dfrac{}{xP}$ and $\dfrac{}{bQ}$

- $\dfrac{xQ}{aP}$ and $\dfrac{}{bQ}$

- $\dfrac{xQ}{xP}$ and $\dfrac{}{bQ}$

- $\dfrac{}{x \xrightarrow{c} y}$ and $\emptyset$

- $\emptyset$ and $\dfrac{}{bQ}$ $\dfrac{xQ}{xP}$

We proceed to present a syntactic format for TSSs from [93, 116, 191], which ensures that an extended TSS is conservative over the original TSS. The following definition is crucial in the formulation of this format.

Definition B.5.2 (Source-dependency). *The* source-dependent *variables in a transition rule ρ are defined inductively as follows:*

- *all variables in the source of ρ are source-dependent;*
- *if $t \xrightarrow{a} t'$ is a premise of ρ and all variables in t are source-dependent, then all variables in t' are source-dependent.*

A transition rule is source-dependent *if all its variables are. A TSS is source-dependent if all its rules are.*

Exercise B.5.2. Say for each transition rule in the consecutive TSSs T_0 in Exercise B.5.1 whether it is source-dependent.

Definition B.5.3 (Freshness). *Let T_0 and T_1 be TSSs over signatures Σ_0 and Σ_1, respectively. A term in $\mathbb{T}(\Sigma_0 \oplus \Sigma_1)$ is said to be* fresh *if it contains a function symbol from $\Sigma_1 \backslash \Sigma_0$. Similarly, a transition label or predicate symbol in T_1 is* fresh *if it does not occur in T_0.*

Exercise B.5.3. Say for each transition rule in the consecutive TSSs T_1 in Exercise B.5.1 whether its source is fresh.

Theorem B.5.1. *Let T_0 and T_1 be TSSs over signatures Σ_0 and Σ_1, respectively, where T_0 and $T_0 \oplus T_1$ are positive after reduction. Under the following conditions, $T_0 \oplus T_1$ is a conservative extension of T_0.*

1. *T_0 is source-dependent.*
2. *For each $\rho \in T_1$,*
   - *either the source of ρ is fresh,*
   - *or ρ has a premise of the form $t \stackrel{a}{\to} t'$ or tP, where*
     - *$t \in \mathbb{T}(\Sigma_0)$;*
     - *all variables in t occur in the source of ρ; and*
     - *t', a, or P is fresh.*

Proof. See [93].

Exercise B.5.4. Which of the extensions of TSSs in Exercise B.5.1 satisfy the restrictions of Theorem B.5.1?

Exercise B.5.5. Let the signature consist of a constant a and a unary function symbol f. Verify that the TSS

$$\frac{}{a \stackrel{c}{\to} a} \qquad \frac{x \stackrel{c}{\to} y}{f(x) \stackrel{c}{\to} f(y)}$$

is source-dependent. What is the LTS generated by this TSS?

Extend the signature with the constant b, and the TSS with the transition rule

$$\frac{}{b \stackrel{c}{\to} b}$$

(which has the fresh constant b as its source). What is the LTS generated by the extended TSS? Verify that the extension is conservative.

B.6 Modal Logics

A variety of so-called *modal logics* (see [131]) have been developed to express properties of LTSs. Modal logic aims to formulate properties of states, and to identify states that satisfy the same properties. Typically, one wants to determine whether a certain predicate holds in state s, or whether there is a transition $s \xrightarrow{a} s'$ such that a certain formula holds in state s'. It can moreover be desirable to express temporal properties, such as that a certain formula will eventually hold, after zero or more transitions, or that a certain formula will never hold. Two states can be distinguished by a modal logic if they do not satisfy the same formulas over this logic. The computation whether a state in an LTS satisfies a modal formula is referred to as *model checking*. Efficient model-checking algorithms have been developed for a range of modal logics. Section 6.4 mentions several tools that have benefited from this algorithmic development. We proceed to present brief descriptions of some modal and temporal logics; see [84, 182] for surveys of these topics.

In the remainder of this section, the unary negation operator $\neg$ and the binary conjunction operator $\wedge$ from boolean logic have their usual meanings: a formula $\neg\phi$ is true if and only if ϕ is false, and a formula $\phi \wedge \phi'$ is true if and only if both ϕ and ϕ' are true. Standard operators such as the binary disjunction operator $\vee$ and the binary implication operator $\rightarrow$ can be expressed using only $\neg$ and $\wedge$: $\phi \vee \phi'$ is true if and only if $\neg(\neg\phi \wedge \neg\phi')$ is true, and $\phi \rightarrow \phi'$ is true if and only if $\neg(\phi \wedge \neg\phi')$ is true.

As before, A denotes a finite set of transition labels, and we assume a finite set of predicates on states, including a special predicate that holds in all states.

Hennessy-Milner Logic. Hennessy-Milner logic (HML) [121], extended with predicates, is a modal logic of which the syntax is defined by the following BNF grammar [9]:

$$\phi \quad ::= \quad P \mid \neg\phi \mid \phi \wedge \phi' \mid \langle a \rangle \phi$$

where a ranges over A and P over the set of predicates. Intuitively, a formula $\langle a \rangle \phi$ expresses that there is an a-transition from the current state to a state where the formula ϕ holds.

Assume an LTS. The states s in this LTS that satisfy an HML formula ϕ, denoted by $s \models \phi$, are defined inductively as follows:

$$
\begin{aligned}
&s \models P && \text{if } sP \\
&s \models \neg\phi && \text{if } s \not\models \phi \\
&s \models \phi \wedge \phi' && \text{if } s \models \phi \text{ and } s \models \phi' \\
&s \models \langle a \rangle \phi && \text{if there is a state } s' \text{ with } s \xrightarrow{a} s' \text{ and } s' \models \phi
\end{aligned}
$$

Exercise B.6.1. Let $A \triangleq \{a, b, c\}$. Give an HML formula that holds for state s_0 in the LTS $\{s_0 \xrightarrow{a} s_1,\ s_1 \xrightarrow{b} s_2,\ s_1 \xrightarrow{c} s_2,\ s_2 P\}$, but not for state s_3 in the LTS $\{s_3 \xrightarrow{a} s_4,\ s_3 \xrightarrow{a} s_5,\ s_4 \xrightarrow{b} s_6,\ s_5 \xrightarrow{c} s_6,\ s_6 P\}$.

The following theorem from [121] states that HML distinguishes regular processes (see Definition B.3.1) up to bisimulation equivalence.

Theorem B.6.1. *Two regular processes p and q are bisimilar if and only if their root states satisfy exactly the same HML formulas.*

By adapting the meaning of formulas $\langle a \rangle \phi'$ and P in HML, one can obtain a modal logic that distinguishes regular processes up to branching bisimulation equivalence; see [82].

We proceed to discuss some existing temporal logics. The modal μ-calculus [139] extends HML with least and greatest fixed-point operators $\nu Z.\phi$ and $\mu Z.\phi$, to express temporal properties. Hennessy and Stirling [122] added a relativised past tense operator to HML, to upgrade it to a temporal logic.

Computation Tree Logic. Computation tree logic (CTL*) [86] is a temporal logic to express properties of unlabelled transition systems. ACTL* [81] is an extension of CTL* to LTSs; its syntax is defined by the following BNF grammar:

$$\phi ::= P \mid \neg \phi \mid \phi \wedge \phi' \mid \langle a \rangle \phi \mid \phi \cup \phi' \mid \mathsf{E}\phi$$

where a ranges over A and P over the set of predicates. Intuitively, $\phi \cup \phi'$ denotes that there is a sequence of transitions from the current state that only visits states in which ϕ holds, until it visits a state in which ϕ' holds. Furthermore, $\mathsf{E}\phi$ expresses that there is a sequence of transitions from the current state, which cannot be extended to a longer sequence, such that the sequence only visits states in which ϕ holds.

Assume an LTS. A *full path* is either an infinite sequence of transitions $s_0 \xrightarrow{a_0} s_1 \xrightarrow{a_1} s_2 \xrightarrow{a_2} \cdots$, or a finite sequence of transitions $s_0 \xrightarrow{a_0} \cdots \xrightarrow{a_{\ell-1}} s_\ell$ where there is no transition $s_\ell \xrightarrow{b} s$. The states s_0 in this LTS that satisfy an ACTL* formula ϕ, denoted by $s_0 \models \phi$, are defined inductively as follows:

$s_0 \models P$ if $s_0 P$

$s_0 \models \neg \phi$ if $s_0 \not\models \phi$

$s_0 \models \phi \wedge \phi'$ if $s_0 \models \phi$ and $s_0 \models \phi'$

$s_0 \models \langle a \rangle \phi$ if there is a state s_1 with $s_0 \xrightarrow{a} s_1$ and $s_1 \models \phi$

$s_0 \models \phi \cup \phi'$ if there is a path $s_0 \xrightarrow{a_0} \cdots \xrightarrow{a_{\ell-1}} s_\ell$ with $s_k \models \phi$ for
                 $k \in \{0, \ldots, \ell - 1\}$ and $s_\ell \models \phi'$

$s_0 \models \mathsf{E}\phi$ if there is a full path, starting in s_0, such that $s \models \phi$ for all
                 states s on this full path

Model checking of ACTL* formulas is PSPACE-complete [87, 181]. In order to obtain a fragment of ACTL* on which model checking is feasible, we make a distinction between so-called state and path formulas:

- each predicate P is a state formula;
- if ϕ and ϕ' are state formulas, then $\phi \wedge \phi'$ is a state formula;

- if ϕ is a state or path formula, then $\neg\phi$ is a state or path formula, respectively;
- if ϕ is a state formula, then $\langle a \rangle \phi$ is a path formula;
- if ϕ and ϕ' are state formulas, then $\phi \cup \phi'$ is a path formula;
- if ϕ is a path formula, then $E \phi$ is a state formula.

ACTL [70, 81] consists of the set of state formulas that are thus defined. An efficient model-checking algorithm exists for ACTL [71, 81], which is linear both in the size of the LTS and in the size of the formula. ACTL is usually referred to as a *branching-time* temporal logic, because the operator $E \phi$ constitutes an explicit quantification over full paths.

Linear Temporal Logic. Linear temporal logic (LTL) [172], here presented with transition labels, is defined by the following BNF grammar:

$$\phi \quad ::= \quad P \mid \neg\phi \mid \phi \wedge \phi' \mid \langle a \rangle \phi \mid \phi \cup \phi'.$$

The model-checking algorithm for LTL [142] is linear in the size of the LTS, but exponential in the size of the formula. From a practical point of view this exponential complexity need not be problematic, because in general the size of a formula is small with respect to the size of the LTS against which it is checked. LTL is referred to as a *linear-time* temporal logic, because formulas are interpreted over linear sequences of states. See [85] for a comparison of branching-time and linear-time temporal logics.

Solutions to Selected Exercises

2.1.1 $a \cdot (b + c)$ and $(a \cdot b) + (a \cdot c)$.

2.2.1 $((a+b)\cdot(a+c))\cdot d \xrightarrow{a} (a+c)\cdot d$, $((a+b)\cdot(a+c))\cdot d \xrightarrow{b} (a+c)\cdot d$, $(a+c)\cdot d \xrightarrow{a} d$, $(a+c)\cdot d \xrightarrow{c} d$, and $d \xrightarrow{d} \sqrt{}$.

2.3.1
- yes: $(b+c)a + ba + ca \, \mathcal{B} \, ba + ca$ and $a \, \mathcal{B} \, a$;
- no;
- yes: $(a+a)(bc)+(ab)(c+c) \, \mathcal{B} \, (a(b+b))(c+c)$, $bc \, \mathcal{B} \, (b+b)(c+c)$, $b(c+c) \, \mathcal{B} \, (b+b)(c+c)$, $cBc+c$, and $c+cBc+c$.

2.3.3 Base case: $a \xrightarrow{a} \sqrt{}$, while aa cannot terminate successfully by the execution of an a-transition. Hence, $a \not\leftrightarrow aa$.

Inductive case: $a^{k+1} \xrightarrow{a} a^k$ is the only transition of a^{k+1}, while $a^{k+2} \xrightarrow{a} a^{k+1}$ is the only transition of a^{k+2}. By induction, a^k and a^{k+1} cannot be related by a bisimulation relation. Hence, $a^{k+1} \not\leftrightarrow a^{k+2}$.

2.4.1 The crux of this exercise is to show that A2' and A3 together prove A1.
$$x + y \stackrel{A3}{=} (x+y)+(x+y) \stackrel{A2'}{=} y+((x+y)+x) \stackrel{A2'}{=} y+(y+(x+x)) \stackrel{A2'}{=}$$
$$((x+x)+y)+y \stackrel{A2'}{=} (x+(y+x))+y \stackrel{A2'}{=} (y+x)+(y+x) \stackrel{A3}{=} y+x.$$

2.4.3 Let $s + t \, \mathcal{B} \, t + s$ and $v \, \mathcal{B} \, v$ for all basic process terms v. It is easy to see that $s + t \xrightarrow{a} w$ or $s + t \xrightarrow{a} \sqrt{}$ if and only if $t + s \xrightarrow{a} w$ or $t + s \xrightarrow{a} \sqrt{}$, respectively.

Let $(s + t)u \, \mathcal{B} \, su + tu$ and $v \, \mathcal{B} \, v$ for all basic process terms v. It is easy to see that $(s + t)u \xrightarrow{a} w$ or $(s + t)u \xrightarrow{a} \sqrt{}$ if and only if $su + tu \xrightarrow{a} w$ or $su + tu \xrightarrow{a} \sqrt{}$, respectively.

Let $(vt)u \, \mathcal{B} \, v(tu)$ and $v \, \mathcal{B} \, v$ for all basic process terms v. If $v \xrightarrow{a} v'$, then $(vt)u \xrightarrow{a} (v't)u$ corresponds to $v(tu) \xrightarrow{a} v'(tu)$; if $v \xrightarrow{a} \sqrt{}$, then $(vt)u \xrightarrow{a} tu$ corresponds to $v(tu) \xrightarrow{a} tu$.

2.4.5 $(ab)c \to a(bc)$, while both $(ab)c$ and $a(bc)$ have weight 8.

2.4.6
- The rewrite rules for BPA reduce $((a+a)(b+b))(c+c)$ to the normal form $a(bc)$.
- The rewrite rules for BPA reduce both $(a+a)(bc)+(ab)(c+c)$ and $(a(b+b))(c+c)$ to the normal form $a(bc)$.
- The rewrite rules for BPA reduce $((a+b)c+ac)d$ to the normal form $a(cd)+b(cd)$, and $(b+a)(cd)$ to the normal form $b(cd) + a(cd)$. These two normal forms are equivalent modulo AC of the $+$.

3.3.2 $(a+b)\|c \leftrightarrow (a+b)c+c(a+b)+\gamma(a,c)+\gamma(b,c)$ is not bisimilar with $a\|c+b\|c \leftrightarrow ac+ca+\gamma(a,c)+bc+cb+\gamma(b,c)$.
$a\mathbb{L}(b+c) \leftrightarrow a(b+c)$ is not bisimilar with $a\mathbb{L}b+a\mathbb{L}c \leftrightarrow ab+ac$.

3.3.6 $a\|((b+c)d) \overset{M1}{=} a\mathbb{L}((b+c)d) + ((b+c)d)\mathbb{L}a + a|((b+c)d) \overset{A4}{=} a\mathbb{L}((b+c)d) +$
$((b+c)d)\mathbb{L}a + a|(bd+cd) \overset{CM10}{=} a\mathbb{L}((b+c)d) + ((b+c)d)\mathbb{L}a + a|(bd) + a|(cd) \overset{CM6}{=}$
$a\mathbb{L}((b+c)d) + ((b+c)d)\mathbb{L}a + \gamma(a,b)d + \gamma(a,c)d \overset{CM7}{=} a\mathbb{L}((b+c)d) + ((b+c)d)\mathbb{L}a +$
$(bd)|a + (cd)|a \overset{CM9}{=} a\mathbb{L}((b+c)d) + ((b+c)d)\mathbb{L}a + (bd+cd)|a \overset{A4}{=} a\mathbb{L}((b+c)d) +$
$((b+c)d)\mathbb{L}a + ((b+c)d)|a \overset{M1}{=} ((b+c)d)\|a.$

3.3.7 Sketch: First prove that $s|t = t|s$ holds for basic process terms s and t, by
induction with respect to their sizes. Next, observe that each process term in
PAP is provably equal to a basic process term.
$s\|t \overset{M1}{=} (s\mathbb{L}t + t\mathbb{L}s) + s|t \overset{A1}{=} (t\mathbb{L}s + s\mathbb{L}t) + s|t = (t\mathbb{L}s + s\mathbb{L}t) + t|s \overset{M1}{=} t\|s.$

3.3.8 $a\mathbb{L}b \underleftrightarrow{} ab$ is not bisimilar with $b\mathbb{L}a \underleftrightarrow{} ba$.
$(a\mathbb{L}b)\mathbb{L}c \underleftrightarrow{} a(bc + cb + \gamma(b,c))$ is not bisimilar with $a\mathbb{L}(b\mathbb{L}c) \underleftrightarrow{} a(bc)$.

3.4.2 $\partial_{\{a\}}(ac);$
$\partial_{\{a\}}((a+b)c) \overset{b}{\to} \partial_{\{a\}}(c) \overset{c}{\to} \sqrt{};$
$\partial_{\{c\}}((a+b)c) \overset{a}{\to} \partial_{\{c\}}(c)$ and $\partial_{\{c\}}((a+b)c) \overset{b}{\to} \partial_{\{c\}}(c);$
$\partial_{\{a,b\}}((ab)\|(ba)) \overset{c}{\to} \partial_{\{a,b\}}(b\|a) \overset{c}{\to} \sqrt{}.$

3.4.5 yes; no; yes; yes; no.

3.4.7 Let $\gamma(a,b) \overset{\Delta}{=} c$. Then $\partial_{\{a,b\}}(a\|b) \underleftrightarrow{} c$, while $\partial_{\{a,b\}}(a)\|\partial_{\{a,b\}}(b) \underleftrightarrow{} \delta$.

3.4.9 $\delta = s + t \overset{A3}{=} (s+t) + (s+t) \overset{A1;2}{=} s + (s + (t+t)) \overset{A3}{=} s + (s+t) = s + \delta \overset{A6}{=} s.$

3.4.10
- $\delta\|a \overset{M1}{=} \delta\mathbb{L}a + a\mathbb{L}\delta + \delta|a \overset{LM2,11,CM12}{=} \delta a + a\delta + \delta \overset{A6,7}{=} a\delta.$

$$
\begin{array}{ll}
\partial_{\{a,b\}}((ab)\|(ba)) & \overset{M1}{=} \partial_{\{a,b\}}((ab)\mathbb{L}(ba) + (ba)\mathbb{L}(ab) + (ab)|(ba)) \\
& \overset{LM3,CM8}{=} \partial_{\{a,b\}}(a(b\|(ba)) + b(a\|(ab)) + c(b\|a)) \\
& \overset{D1,2,4,5}{=} \delta\partial_{\{a,b\}}(b\|(ba)) + \delta\partial_{\{a,b\}}(a\|(ab)) + c\partial_{\{a,b\}}(b\|a) \\
& \overset{A6,7}{=} c\partial_{\{a,b\}}(b\|a) \\
& \overset{M1}{=} c\partial_{\{a,b\}}(b\mathbb{L}a + a\mathbb{L}b + b|a) \\
& \overset{LM2,CM5}{=} c\partial_{\{a,b\}}(ba + ab + c) \\
& \overset{D1,2,4,5}{=} c(\delta\partial_{\{a,b\}}(a) + \delta\partial_{\{a,b\}}(b) + c) \\
& \overset{A6,7}{=} cc.
\end{array}
$$

- $send(d)$, $read(d)$, and $comm(d)$ are abbreviated to $s(d)$, $r(d)$, and $c(d)$, respectively, for $d \in \{0,1\}$, and H denotes $\{s(0), s(1), r(0), r(1)\}$.

$$
\begin{array}{ll}
& (s(0) + s(1))\|(r(0) + r(1)) \\
\overset{M1}{=} & (s(0) + s(1))\mathbb{L}(r(0) + r(1)) + (r(0) + r(1))\mathbb{L}(s(0) + s(1)) \\
& + (s(0) + s(1))|(r(0) + r(1)) \\
\overset{LM4,CM9,10}{=} & s(0)\mathbb{L}(r(0) + r(1)) + s(1)\mathbb{L}(r(0) + r(1)) + r(0)\mathbb{L}(s(0) + s(1)) \\
& + r(1)\mathbb{L}(s(0) + s(1)) + s(0)|r(0) + s(0)|r(1) + s(1)|r(0) \\
& + s(1)|r(1) \\
\overset{LM2,CM5}{=} & s(0)(r(0) + r(1)) + s(1)(r(0) + r(1)) + r(0)(s(0) + s(1)) \\
& + r(1)(s(0) + s(1)) + c(0) + \delta + \delta + c(1) \\
\overset{A6}{=} & s(0)(r(0) + r(1)) + s(1)(r(0) + r(1)) + r(0)(s(0) + s(1)) \\
& + r(1)(s(0) + s(1)) + c(0) + c(1).
\end{array}
$$

Hence,

$$
\begin{aligned}
&\partial_H((s(0) + s(1))\|(r(0) + r(1)))\\
={}& \partial_H(s(0)(r(0) + r(1)) + s(1)(r(0) + r(1)) + r(0)(s(0) + s(1))\\
&+ r(1)(s(0) + s(1)) + c(0) + c(1))\\
\overset{D1,2,4,5}{=}{}& \delta\partial_H(r(0) + r(1)) + \delta\partial_H(r(0) + r(1)) + \delta\partial_H(s(0) + s(1))\\
&+ \delta\partial_H(s(0) + s(1)) + c(0) + c(1)\\
\overset{A6,7}{=}{}& c(0) + c(1).
\end{aligned}
$$

3.4.11

ALT1	$alt(a, x) = a \cdot x$
ALT2	$alt(\delta, x) = \delta$
ALT3	$alt(a \cdot x, y) = a \cdot alt(y, x)$
ALT4	$alt(x + y, z) = alt(x, z) + alt(y, z)$

The first axiom says that if the left-hand side s of $alt(s, t)$ terminates successfully by the execution of a, then $alt(s, t)$ executes a after which it results to t.

The second axiom says that if the left-hand side s of $alt(s, t)$ cannot execute any actions, then $alt(s, t)$ cannot execute any actions.

The third axiom says that if the left-hand side s of $alt(s, t)$ executes a to evolve into s', then $alt(s, t)$ executes a to evolve into $alt(t, s')$.

The fourth axiom says that if the left-hand side s of $alt(s, t)$ can execute actions a_i to evolve into s_i, then $alt(s, t)$ can execute the a_i to evolve into $alt(t, s_i)$.

Direct the axioms for alt from left to right, and add them to the TRS for ACP. An appropriate weight function shows that the resulting TRS is terminating modulo AC of the $+$. One can show that normal forms do not contain occurrences of alt. Let the process terms s and t in ACP extended with alt be bisimilar. Then one can apply the rewrite rules to remove all occurrences of $\|$, $\mathbb{L}$, $|$, ∂_H, and alt from s and t, to obtain normal forms n and n', respectively, in BPA extended with δ. Since the axioms are sound, n and n' are bisimilar. This implies $n =_{AC} n'$, so $s = n =_{AC} n' = t$.

4.1.1 Let $\gamma(a, a) \overset{\Delta}{=} \delta$ and $\gamma(a, b) \overset{\Delta}{=} \delta$. Then $aaa \cdots$ and $(a + b)(a + b)(a + b) \cdots$ are two non-bisimilar solutions.

4.1.2
- $\{X = aY, Y = bX\}$ is already in the desired form.
- The right-hand side of $Y = aX$ is already in the desired form. The right-hand side of $X = Y$ is brought into the desired form by replacing Y by the right-hand side aX of its recursive equation.
- The right-hand side of $X = (a + b) \mathbb{L} X$ is brought into the desired form by applications of axioms LM2,4: $(a + b) \mathbb{L} X = aX + bX$.

4.1.3 $Z = bZ$ is already in the desired form.

$Y = Z + a$ is brought into the desired form by replacing Z by the right-hand side bZ of its recursive equation.

$X = Y \| Z$ is brought into the desired form by replacing Y and Z by $bZ + a$ and bZ, respectively, and manipulating the resulting term $(bZ + a) \| (bZ)$ by the axioms: $(bZ + a) \| (bZ) = (bZ + a) \mathbb{L} (bZ) + (bZ) \mathbb{L} (bZ + a) + (bZ + a)|(bZ) = (bZ) \mathbb{L} (bZ) + a \mathbb{L} (bZ) + (bZ) \mathbb{L} (bZ + a) + (bZ)|(bZ) + a|(bZ) = b(Z\|(bZ)) + a(bZ) + b(Z\|(bZ + a)) + c(Z\|Z) + cZ$.

4.2.1

$$b \xrightarrow{b} \checkmark \qquad (\frac{}{v \xrightarrow{v} \checkmark}, \qquad v := b)$$

$$b\langle X|E\rangle \xrightarrow{b} \langle X|E\rangle \quad (\frac{x \xrightarrow{v} \checkmark}{xy \xrightarrow{v} y}, \qquad v := b, \; x := b, \; y := \langle X|E\rangle)$$

$$\langle Y|E\rangle \xrightarrow{b} \langle X|E\rangle \quad (\frac{b\langle X|E\rangle \xrightarrow{v} y}{\langle Y|E\rangle \xrightarrow{v} y}, \; v := b, \; y := \langle X|E\rangle)$$

4.2.2
- $\langle X \mid X=ab\rangle \xrightarrow{a} b \xrightarrow{b} \checkmark$;
- $\langle X \mid X=YX, Y=bY\rangle \xrightarrow{b} \langle Y \mid X=YX, Y=bY\rangle\langle X \mid X=YX, Y=bY\rangle$,
  $\langle Y \mid X=YX, Y=bY\rangle\langle X \mid X=YX, Y=bY\rangle \xrightarrow{b}$
  $\qquad\qquad\qquad \langle Y \mid X=YX, Y=bY\rangle\langle X \mid X=YX, Y=bY\rangle$;
- $\langle X \mid X=aXb\rangle b^k \xrightarrow{a} \langle X \mid X=aXb\rangle b^{k+1}$ for $k \in \mathbb{N}$;
- $\langle X \mid X=aXb+c\rangle b^k \xrightarrow{a} \langle X \mid X=aXb+c\rangle b^{k+1}$ for $k \in \mathbb{N}$,
  $\langle X \mid X=aXb+c\rangle \xrightarrow{c} \checkmark$,
  $\langle X \mid X=aXb+c\rangle b^{k+1} \xrightarrow{c} b^{k+1}$ for $k \in \mathbb{N}$.

4.2.3 Let $s_{k,\ell}$ represent the state in which there are k zeros and ℓ ones in the bag (so the root state is $s_{0,0}$). Let $s_{k,\ell} B(\cdots(((\langle X|E\rangle\|out(d_1))\|out(d_2)\|\cdots)\|out(d_n)$ if the sequence $d_1\cdots d_n$ of elements in $\{0,1\}$ contains k zeros and ℓ ones.

4.2.4 Assume recursion variables $X_{k,\ell}$ for $k, \ell \in \mathbb{N}$. Intuitively, $X_{k,\ell}$ represents the state in which there are k zeros and ℓ ones in the bag. This is captured by the following recursive equations, where k and ℓ range over $\mathbb{N}$:

$$
\begin{aligned}
X_{0,0} &= in(0)X_{1,0} + in(1)X_{0,1}\\
X_{0,\ell+1} &= in(0)X_{1,\ell+1} + in(1)X_{0,\ell+2} + out(1)X_{0,\ell}\\
X_{k+1,0} &= in(0)X_{k+2,0} + in(1)X_{k+1,1} + out(0)X_{k,0}\\
X_{k+1,\ell+1} &= in(0)X_{k+2,\ell+1} + in(1)X_{k+1,\ell+2} + out(0)X_{k,\ell+1} + out(1)X_{k+1,\ell}.
\end{aligned}
$$

4.3.1 Since $t = t$, RSP yields $t = \langle X \mid X=X\rangle$ for all process terms t.

4.3.2 $E \triangleq \{X = aX\}$. $\langle X|E\rangle \mathbin{\rotatebox[origin=c]{180}{L}} b \xrightarrow{a} \langle X|E\rangle\| b \xrightarrow{b} \langle X|E\rangle$, while $\langle X|E\rangle b \leftrightarrow \langle X|E\rangle$.

4.3.3
- $\langle X \mid X=aX+b\rangle \overset{\mathrm{RDP}}{=} a\langle X \mid X=aX+b\rangle+b$. So the desired equation follows by RSP.
- $\langle X \mid X=aX\rangle \overset{\mathrm{RDP}}{=} a\langle X \mid X=aX\rangle$. So the desired equation follows by RSP.
- $\langle Z \mid Z=aZ\rangle \overset{\mathrm{RDP}}{=} a\langle Z \mid Z=aZ\rangle \overset{\mathrm{RDP}}{=} aa\langle Z \mid Z=aZ\rangle$. By RSP,

$$\langle Z \mid Z=aZ\rangle = \langle X \mid X=aaX\rangle.$$

  $\langle Z \mid Z=aZ\rangle \overset{\mathrm{RDP}}{=} a\langle Z \mid Z=aZ\rangle \overset{\mathrm{RDP}}{=} aa\langle Z \mid Z=aZ\rangle \overset{\mathrm{RDP}}{=} aaa\langle Z \mid Z=aZ\rangle$. By RSP,

$$\langle Z \mid Z=aZ\rangle = \langle Y \mid Y=aaaY\rangle.$$

$$
\begin{aligned}
\langle Z \mid Z=(a+b)Z\rangle &\overset{\mathrm{RDP}}{=} (a+b)\langle Z \mid Z=(a+b)Z\rangle\\
&\overset{A4}{=} a\langle Z \mid Z=(a+b)Z\rangle + b\langle Z \mid Z=(a+b)Z\rangle\\
&\overset{\mathrm{RDP}}{=} a\langle Z \mid Z=(a+b)Z\rangle + b(a+b)\langle Z \mid Z=(a+b)Z\rangle.
\end{aligned}
$$

So by RSP, $\langle Z \mid Z=(a+b)Z\rangle = \langle X \mid X=aX+b(a+b)X\rangle$.
Likewise it can be derived that $\langle Z \mid Z=(a+b)Z\rangle = \langle Y \mid Y=bY+a(a+b)Y\rangle$.
- By $\mathcal{E}_{\text{ACP}}$, RDP, and commutativity of the merge,

$$\langle X \mid X=aX\rangle \| \langle Y \mid Y=bY\rangle = (a + b + \gamma(a,b))((\langle X \mid X=aX\rangle \| \langle Y \mid Y=bY\rangle)).$$

So by RSP, $\langle X \mid X=aX\rangle \| \langle Y \mid Y=bY\rangle = \langle Z \mid Z=(a+b+\gamma(a,b))Z\rangle$.
- By RDP and A4,

$$\langle X \mid X=aX+b\rangle\langle Y \mid Y=(a+b)Y\rangle$$
$$= a\langle X \mid X=aX+b\rangle\langle Y \mid Y=(a+b)Y\rangle + b\langle Y \mid Y=(a+b)Y\rangle$$

and by RDP, $\langle Y \mid Y=(a+b)Y\rangle = (a + b)\langle Y \mid Y=(a+b)Y\rangle$.
So by RSP, $\langle X \mid X=aX+b\rangle\langle Y \mid Y=(a+b)Y\rangle = \langle V \mid V=aV+bW, W=(a+b)W\rangle$.
Furthermore, by RDP and A4,

$$\langle Z \mid Z=(a+b)Z\rangle = a\langle Z \mid Z=(a+b)Z\rangle + b\langle Z \mid Z=(a+b)Z\rangle,$$

and by RDP, $\langle Z \mid Z=(a+b)Z\rangle = (a + b)\langle Z \mid Z=(a+b)Z\rangle$.
So by RSP, $\langle Z \mid Z=(a+b)Z\rangle = \langle V \mid V=aV+bW, W=(a+b)W\rangle$.
- $\langle X \mid X=aX\rangle b \overset{\text{RDP}}{=} a\langle X \mid X=aX\rangle b$, so by RSP $\langle X \mid X=aX\rangle b = \langle X \mid X=aX\rangle$.
- By the previous item, $\langle X \mid X=aX\rangle = \langle X \mid X=aX\rangle b \overset{\text{RDP}}{=} a\langle X \mid X=aX\rangle b$.
Hence, by RSP, $\langle X \mid X=aX\rangle = \langle Y \mid Y=aYb\rangle$.

4.3.4 Let E consist of the recursive equations $X_1 = a(X_2b + c)$, $X_2 = cX_2 + bX_3$, and $X_3 = a(X_1 + X_3)X_2$. It is easy to see that E is guarded. So RSP yields $t_i = \langle X_i | E\rangle$ for $i \in \{1, 2, 3\}$.

4.3.5 Since $t_1 = at_2$ and $t_2 = at_1$, substituting t_1 for Y_1 and t_2 for Y_2 is a solution for $E \triangleq \{Y_1=aY_2, Y_2=aY_1\}$. So by RSP, $t_1 = \langle Y_1|E\rangle$.
RDP yields $\langle X \mid X=aX\rangle = a\langle X \mid X=aX\rangle$, so substituting $\langle X \mid X=aX\rangle$ for Y_1 and Y_2 is a solution for $\{Y_1=aY_2, Y_2=aY_1\}$. Hence, by RSP, $\langle X \mid X=aX\rangle = \langle Y_1|E\rangle$.

4.4.1 $\langle X \mid X=aX+bY, Y=cX + aY\rangle$.

4.5.3 Apply induction on n. The case $n \equiv 0$ is trivial; let $n > 0$.
If $\pi_n(s) \overset{a}{\to} \sqrt{}$, then $\pi_{n+1}(s) \overset{a}{\to} \sqrt{}$. Since $\pi_{n+1}(s) \underline{\leftrightarrow} \pi_{n+1}(t)$, this implies $\pi_{n+1}(t) \overset{a}{\to} \sqrt{}$. Since $n > 0$, $\pi_n(t) \overset{a}{\to} \sqrt{}$. Likewise, $\pi_n(t) \overset{a}{\to} \sqrt{}$ implies $\pi_n(s) \overset{a}{\to} \sqrt{}$.
If $\pi_n(s) \overset{a}{\to} \pi_{n-1}(s')$, then $\pi_{n+1}(s) \overset{a}{\to} \pi_n(s')$. Since $\pi_{n+1}(s) \underline{\leftrightarrow} \pi_{n+1}(t)$, this implies $\pi_{n+1}(t) \overset{a}{\to} \pi_n(t')$ with $\pi_n(s') \underline{\leftrightarrow} \pi_n(t')$. Then $\pi_n(t) \overset{a}{\to} \pi_{n-1}(t')$, and by induction $\pi_{n-1}(s') \underline{\leftrightarrow} \pi_{n-1}(t')$. Likewise, $\pi_n(t) \overset{a}{\to} \pi_{n-1}(t')$ implies $\pi_n(s) \overset{a}{\to} \pi_{n-1}(s')$ with $\pi_{n-1}(t') \underline{\leftrightarrow} \pi_{n-1}(s')$.

4.5.4 $S_n \triangleq \{n, n + 1, n + 2, \ldots\}$ for $n \in \mathbb{N}$.

4.5.5 Consider the process graphs $\{s \overset{a}{\to} s', s' \overset{a}{\to} s'\} \cup \{s \overset{a}{\to} s_n, s_{n+1} \overset{a}{\to} s_n \mid n \in \mathbb{N}\}$ and $\{\hat{s} \overset{a}{\to} \hat{s}_n, \hat{s}_{n+1} \overset{a}{\to} \hat{s}_n \mid n \in \mathbb{N}\}$, with root states s and $\hat{s}$, respectively. s and $\hat{s}$ are bisimilar up to any finite depth, but s has an infinite trace of a-transitions $(s \overset{a}{\to} s' \overset{a}{\to} s' \overset{a}{\to} \cdots)$ while $\hat{s}$ has no such trace. So s and $\hat{s}$ are not bisimilar.

4.5.6 We derive for $k, n \in \mathbb{N}$, by induction on n:

$$\pi_n(\langle X \mid X=aXb+b\rangle b^k) = \pi_n(\langle Y \mid Y=aZb+b, Z=aYb+b\rangle b^k)$$
$$\pi_n(\langle X \mid X=aXb+b\rangle b^k) = \pi_n(\langle Z \mid Y=aZb+b, Z=aYb+b\rangle b^k)$$

(The desired equality then follows by AIP, taking $k \equiv 0$.)
The base case $n \equiv 0$ is trivial. Using induction one can derive:

$$\pi_{n+1}(\langle X \mid X=aXb+b\rangle b^k)$$

$$\overset{RDP,A4}{=} \pi_{n+1}(a\langle X \mid X=aXb+b\rangle b^{k+1} + b^{k+1})$$

$$\overset{PR1\text{-}3}{=} a\pi_n(\langle X \mid X=aXb+b\rangle b^{k+1}) + \pi_{n+1}(b^{k+1})$$

$$= a\pi_n(\langle Z \mid Y=aZb+b, Z=aYb+b\rangle b^{k+1}) + \pi_{n+1}(b^{k+1})$$

$$\overset{PR1\text{-}3}{=} \pi_{n+1}(a\langle Z \mid Y=aZb+b, Z=aYb+b\rangle b^{k+1} + b^{k+1})$$

$$\overset{RDP,A4}{=} \pi_{n+1}(\langle Y \mid Y=aZb+b, Z=aYb+b\rangle b^k).$$

Likewise one can derive

$$\pi_{n+1}(\langle X \mid X=aXb+b\rangle b^k) = \pi_{n+1}(\langle Z \mid Y=aZb+b, Z=aYb+b\rangle b^k).$$

5.1.1 $a\,B_1\,a\tau$, $\sqrt{}\,B_1\,\tau$, and $\sqrt{}\,B_1\,\sqrt{}$ proves $a \underleftrightarrow{}_b a\tau$;
$a\,B_2\,\tau a$, $a\,B_2\,a$, and $\sqrt{}\,B_2\,\sqrt{}$ proves $a \underleftrightarrow{}_b \tau a$.
$a\tau\,B_3\,\tau a$, $a\tau\,B_3\,a$, $\tau\,B_3\,\sqrt{}$, and $\sqrt{}\,B_3\,\sqrt{}$ proves $a\tau \underleftrightarrow{}_b \tau a$.

5.1.2 $\tau(\tau(a+b)+b)+aBa+b$, $\tau(a+b)+bBa+b$, $a+bBa+b$, and $\sqrt{}B\sqrt{}$.

5.1.4 not branching bisimilar; bisimilar; branching bisimilar but not rooted branching bisimilar; rooted branching bisimilar but not bisimilar; not branching bisimilar.

5.2.1

$$\langle X \mid X=aY+\tau Y, Y=bX+\tau X\rangle \overset{\tau}{\to} \langle Y \mid X=aY+\tau Y, Y=bX+\tau X\rangle$$
$$\overset{\tau}{\to} \langle X \mid X=aY+\tau Y, Y=bX+\tau X\rangle.$$

For each $c \in A$, a solution for $\{X=aY+\tau Y, Y=bX+\tau X\}$ is to substitute $(a+\tau)\langle Z \mid Z=aZ+bZ+cZ\rangle$ for X and $(b+\tau)\langle Z \mid Z=aZ+bZ+cZ\rangle$ for Y. For different atomic actions c, the solutions above are not rooted branching bisimilar.

5.2.2

$$\overline{\sqrt{}\downarrow} \qquad\qquad \overline{v \overset{v}{\to} \sqrt{}}$$

$$\frac{x\downarrow}{x+y\downarrow} \qquad \frac{x \overset{v}{\to} x'}{x+y \overset{v}{\to} x'} \qquad \frac{y\downarrow}{x+y\downarrow} \qquad \frac{y \overset{v}{\to} y'}{x+y \overset{v}{\to} y'}$$

$$\frac{x\downarrow \quad y\downarrow}{x\cdot y\downarrow} \qquad \frac{x\downarrow \quad y \overset{v}{\to} y'}{x\cdot y \overset{v}{\to} y'} \qquad \frac{x \overset{v}{\to} x'}{x\cdot y \overset{v}{\to} x'\cdot y}$$

$$\frac{x\downarrow \quad y\downarrow}{x\|y\downarrow} \qquad \frac{x \overset{v}{\to} x'}{x\|y \overset{v}{\to} x'\|y} \qquad \frac{y \overset{v}{\to} y'}{x\|y \overset{v}{\to} x\|y'} \qquad \frac{x \overset{v}{\to} x' \quad y \overset{w}{\to} y'}{x\|y \overset{\gamma(v,w)}{\to} x'\|y'}$$

$$\frac{x \overset{v}{\to} x'}{x\Vert y \overset{v}{\to} x'\|y} \qquad \frac{x\downarrow \quad y\downarrow}{x|y\downarrow} \qquad \frac{x \overset{v}{\to} x' \quad y \overset{w}{\to} y'}{x|y \overset{\gamma(v,w)}{\to} x'\|y'}$$

$$\frac{x\downarrow}{\partial_H(x)\downarrow} \qquad\qquad \frac{x \overset{v}{\to} x'}{\partial_H(x) \overset{v}{\to} \partial_H(x')}\; v \notin H$$

$$\frac{t_i(\langle X_1|E\rangle,\ldots,\langle X_n|E\rangle)\downarrow}{\langle X_i|E\rangle\downarrow} \quad \frac{t_i(\langle X_1|E\rangle,\ldots,\langle X_n|E\rangle) \overset{v}{\to} y}{\langle X_i|E\rangle \overset{v}{\to} y}$$

5.3.2 $\langle X \mid X=aX \rangle$ and $\langle Y \mid Y=bY \rangle$.

5.3.3

- $a(\tau b + b) \stackrel{A3}{=} a(\tau(b+b)+b) \stackrel{B2}{=} a(b+b) \stackrel{A3}{=} ab$.

- $a(\tau(b+c)+b) \stackrel{B2}{=} a(b+c) \stackrel{A1}{=} a(c+b) \stackrel{B2}{=} a(\tau(c+b)+c) \stackrel{A1}{=} a(\tau(b+c)+c)$.

- Since each process term in ACP_τ can be reduced to a normal form in BPA extended with δ and τ, it may be assumed that s is a normal form $\sum_i a_i s_i + \sum_j b_j$, where the s_i are normal forms. The desired equation can be proved by structural induction with respect to the size of the normal form s. By induction we have $a_i(s_i \|(\tau t)) = a_i(s_i \| t)$.

$$
\begin{aligned}
a(s\|(\tau t)) &= a((\sum_i a_i s_i + \sum_j b_j)\|(\tau t)) \\
&= a(\tau(t\|s) + \sum_i a_i(s_i\|(\tau t)) + \sum_j b_j \tau t) \\
&= a(\tau(t\|s) + \sum_i a_i(s_i\| t) + \sum_j b_j t) \\
&= a(\tau(s \mathbin{\llcorner} t + t \mathbin{\llcorner} s) + s \mathbin{\llcorner} t) \\
&= a(s \mathbin{\llcorner} t + t \mathbin{\llcorner} s) \\
&= a(s\| t).
\end{aligned}
$$

$$
\begin{aligned}
\langle X \mid X=aY, Y=\tau X \rangle &\stackrel{RDP}{=} a\langle Y \mid X=aY, Y=\tau X \rangle \\
&\stackrel{RDP}{=} a\tau\langle X \mid X=aY, Y=\tau X \rangle \\
&\stackrel{B1}{=} a\langle X \mid X=aY, Y=\tau X \rangle.
\end{aligned}
$$

So by RSP, $\langle X \mid X=aY, Y=\tau X \rangle = \langle Z \mid Z=aZ \rangle$.

$$
\begin{aligned}
\langle Z \mid Z=(a+b)Z \rangle &\stackrel{RDP}{=} (a+b)(a+b)\langle Z \mid Z=(a+b)Z \rangle \\
&\stackrel{B2,A4}{=} (a+b)(\tau(a+b)+b)\langle Z \mid Z=(a+b)Z \rangle \\
&\stackrel{RDP,A4}{=} (a+b)(\tau+b)\langle Z \mid Z=(a+b)Z \rangle.
\end{aligned}
$$

Hence, substituting $\langle Z \mid Z=(a+b)Z \rangle$ for X and $(\tau+b)\langle Z \mid Z=(a+b)Z \rangle$ for Y is a solution for $\{X=(a+b)Y, Y=(\tau+b)X\}$. So by RSP,

$$\langle Z \mid Z=(a+b)Z \rangle = \langle X \mid X=(a+b)Y, Y=(\tau+b)X \rangle.$$

5.3.4 $\tau \mathbin{\underline{\leftrightarrow}_{rb}} \tau\tau$, but $\pi_1(\tau) = \tau$ and $\pi_1(\tau\tau) = \tau\delta$ are not rooted branching bisimilar. π_{n+1} occurs at the right-hand side of the conclusion of the transition rule for π_{n+2}. Furthermore, in the transition rule for π_{n+1}, the argument x of the source $\pi_{n+1}(x)$ is the left-hand side of the premise. Since there is no patience rule for the argument of π_{n+1}, this combination violates the RBB cool format.

5.3.5 Let a range over A (so $a \neq \tau$).

$$
\frac{x \stackrel{a}{\to} \checkmark}{\pi_{n+1}(x) \stackrel{a}{\to} \checkmark} \qquad
\frac{x \stackrel{a}{\to} x'}{\pi_{n+1}(x) \stackrel{a}{\to} \pi_n(x')} \qquad
\frac{x \stackrel{\tau}{\to} \checkmark}{\pi_n(x) \stackrel{\tau}{\to} \checkmark} \qquad
\frac{x \stackrel{\tau}{\to} x'}{\pi_n(x) \stackrel{\tau}{\to} \pi_n(x')}
$$

PR1	$\pi_n(x+y) = \pi_n(x) + \pi_n(y)$
PR2	$\pi_{n+1}(a \cdot x) = a \cdot \pi_n(x)$
PR3	$\pi_0(a \cdot x) = \delta$
PR4	$\pi_n(\delta) = \delta$
PR5	$\pi_n(\tau) = \tau$
PR6	$\pi_n(\tau \cdot x) = \tau \cdot \pi_n(x)$

5.4.1 $a \xrightarrow{a} \sqrt{}$ implies $aa \xrightarrow{a} a$, and $b \xrightarrow{b} \sqrt{}$ implies $bb \xrightarrow{b} b$;

$\gamma(a,b) \equiv c$, so $(aa)\|(bb) \xrightarrow{c} a\|b$;

$c \notin \{a,b\}$, so $\partial_{\{a,b\}}((aa)\|(bb)) \xrightarrow{c} \partial_{\{a,b\}}(a\|b)$;

$c \in \{c\}$, so $\tau_{\{c\}}(\partial_{\{a,b\}}((aa)\|(bb))) \xrightarrow{\tau} \tau_{\{c\}}(\partial_{\{a,b\}}(a\|b))$.

5.4.2 The process graph of $\tau_{\{a\}}(\langle X \mid X{=}aX\rangle)$ consists of the transition

$$\tau_{\{a\}}(\langle X \mid X{=}aX\rangle) \xrightarrow{\tau} \tau_{\{a\}}(\langle X \mid X{=}aX\rangle).$$

Hence, $\tau_{\{a\}}(\langle X \mid X{=}aX\rangle)\, \mathcal{B}\, \delta$ is a branching bisimulation relation.

5.4.4 $\tau_{\{a\}}(\partial_{\{a\}}(a)) \underline{\leftrightarrow}_{rb} \delta$ while $\partial_{\{a\}}(\tau_{\{a\}}(a)) \underline{\leftrightarrow}_{rb} \tau$.

5.4.5 No. A counter-example is $t_1 \equiv ac$ and $t_2 \equiv \tau c$.

5.4.6

$$\tau_{\{b\}}(\langle X \mid X{=}aY, Y{=}bX\rangle) \overset{\text{RDP}}{=} \tau_{\{b\}}(ab\langle Y \mid X{=}aY, Y{=}bX\rangle)$$
$$\overset{\text{T11,2,5}}{=} a\tau\tau_{\{b\}}(\langle X \mid X{=}aY, Y{=}bX\rangle)$$
$$\overset{\text{B1}}{=} a\tau_{\{b\}}(\langle X \mid X{=}aY, Y{=}bX\rangle).$$

So by RSP, $\tau_{\{b\}}(\langle X \mid X{=}aY, Y{=}bX\rangle) = \langle Z \mid Z{=}aZ\rangle$.

5.6.1

- $\tau_{\{a\}}(\langle X \mid X{=}aX\rangle) \xrightarrow{\tau} \tau_{\{a\}}(\langle X \mid X{=}aX\rangle)$, while $\tau\delta \xrightarrow{\tau} \delta$. So it suffices to prove that $\tau_{\{a\}}(\langle X \mid X{=}aX\rangle) \underline{\leftrightarrow}_b \delta$. This is shown by the following branching bisimulation relation $\mathcal{B}$: $\tau_{\{a\}}(\langle X \mid X{=}aX\rangle)\, \mathcal{B}\, \delta$.

- $\tau_{\{a\}}(\langle X \mid X{=}aX{+}b\rangle) \xrightarrow{\tau} \tau_{\{a\}}(\langle X \mid X{=}aX{+}b\rangle)$ and $\tau_{\{a\}}(\langle X \mid X{=}aX{+}b\rangle) \xrightarrow{b} \sqrt{}$, while $b{+}\tau b \xrightarrow{\tau} b$ and $b{+}\tau b \xrightarrow{b} \sqrt{}$. So it suffices to prove that $\tau_{\{a\}}(\langle X \mid X{=}aX{+}b\rangle) \underline{\leftrightarrow}_b b$. This is shown by the following branching bisimulation relation $\mathcal{B}$: $\tau_{\{a\}}(\langle X \mid X{=}aX{+}b\rangle)\, \mathcal{B}\, b$ and $\sqrt{}\, \mathcal{B}\, \sqrt{}$.

- $\tau\tau_{\{a\}}(\langle X \mid X{=}aY{+}b, Y{=}aX{+}c\rangle) \xrightarrow{\tau} \tau_{\{a\}}(\langle X \mid X{=}aX{+}b, Y{=}aX{+}c\rangle)$, while $\tau(b{+}c) \xrightarrow{\tau} b{+}c$. So it suffices to prove that $\tau_{\{a\}}(\langle X \mid X{=}aX{+}b, Y{=}aX{+}c\rangle) \underline{\leftrightarrow}_b b{+}c$. This is shown by the following branching bisimulation relation $\mathcal{B}$: $\tau_{\{a\}}(\langle X \mid X{=}aX{+}b, Y{=}aX{+}c\rangle)\, \mathcal{B}\, b{+}c$, $\tau_{\{a\}}(\langle Y \mid X{=}aX{+}b, Y{=}aX{+}c\rangle)\, \mathcal{B}\, b{+}c$, and $\sqrt{}\, \mathcal{B}\, \sqrt{}$.

5.6.3

- $\{X\}$ is a cluster for $\{a\}$ in $E_1 \triangleq \{X{=}aX + b\}$, with exit b, so

$$\tau_{\{a\}}(\langle X|E_1\rangle) \overset{\text{RDP,T11-5}}{=} \tau\tau_{\{a\}}(\langle X|E_1\rangle) + b \overset{\text{CFAR}}{=} \tau\tau_{\{a\}}(b) + b.$$

$\{Y, Z\}$ is a cluster for $\{a\}$ in $E_2 \triangleq \{Y{=}aZ{+}b, Z{=}aY\}$, with exit b, so

$$\tau_{\{a\}}(\langle Y|E_2\rangle) \overset{\text{RDP,T11-5}}{=} \tau\tau_{\{a\}}(\langle Z|E_2\rangle) + b \overset{\text{CFAR}}{=} \tau\tau_{\{a\}}(b) + b.$$

Hence, $\tau_{\{a\}}(\langle X|E_1\rangle) = \tau\tau_{\{a\}}(b) + b = \tau_{\{a\}}(\langle Y|E_2\rangle)$.

- $\{X, Y\}$ is a cluster for $\{a\}$ in $E_1 \triangleq \{X{=}aY, Y{=}aX{+}bX\}$, with exit bX, so

$$\tau_{\{a\}}(\langle X|E_1\rangle) \overset{\text{RDP,T12,5}}{=} \tau\tau_{\{a\}}(\langle Y|E_1\rangle) \overset{\text{CFAR}}{=} \tau\tau_{\{a\}}(b\langle X|E_1\rangle).$$

Moreover,

$$\tau_{\{a\}}(b\langle X|E_1\rangle) \overset{\text{T11,5}}{=} b\tau_{\{a\}}(\langle X|E_1\rangle).$$

So substituting $\tau_{\{a\}}(\langle X|E_1\rangle)$ for V and $\tau_{\{a\}}(b\langle X|E_1\rangle)$ for W is a solution for $E_2 \triangleq \{V{=}\tau W, W{=}bV\}$. Hence, by RSP, $\tau_{\{a\}}(\langle X|E_1\rangle) = \langle V|E_2\rangle$.

- $\{X,Y\}$ is a cluster for $\{a\}$ in $E \triangleq \{X=aY+b, Y=aX+c\}$, with exits b and c, so

$$\tau_{\{a\}}(\langle X|E\rangle) \overset{\text{RDP,TI1-5}}{=} \tau\tau_{\{a\}}(\langle Y|E\rangle) + b \overset{\text{CFAR}}{=} \tau(b+c)+b.$$

- $\{X,Y\}$ is a cluster for $\{a\}$ in $E \triangleq \{X=aY+bY, Y=aX+cX\}$, with exits bY and cX, so

$$\tau\tau_{\{a\}}(\langle X|E\rangle) \overset{\text{CFAR}}{=} \tau\tau_{\{a\}}(b\langle Y|E\rangle + c\langle X|E\rangle)$$
$$\overset{\text{TI1,4,5}}{=} \tau(b\tau_{\{a\}}(\langle Y|E\rangle) + c\tau_{\{a\}}(\langle X|E\rangle)).$$

Applications of CFAR and TI1,5 give

$$b\tau_{\{a\}}(\langle Y|E\rangle) = b(b\tau_{\{a\}}(\langle Y|E\rangle) + c\tau_{\{a\}}(\langle X|E\rangle))$$
$$c\tau_{\{a\}}(\langle X|E\rangle) = c(b\tau_{\{a\}}(\langle Y|E\rangle) + c\tau_{\{a\}}(\langle X|E\rangle)).$$

So substituting $b\tau_{\{a\}}(\langle Y|E\rangle) + c\tau_{\{a\}}(\langle X|E\rangle)$ for Z is a solution for $\{Z=bZ+cZ\}$. Then RSP yields $b\tau_{\{a\}}(\langle Y|E\rangle) + c\tau_{\{a\}}(\langle X|E\rangle) = \langle Z \mid Z=bZ+cZ\rangle$. Hence,

$$\tau\tau_{\{a\}}(\langle X|E\rangle) = \tau\langle Z \mid Z=bZ+cZ\rangle.$$

6.1.2 $R_0\|S_0 \overset{r_A(d)}{\to} R_0\|T_{d0} \overset{c_B(\bot)}{\to} Q_1\|U_{d0} \overset{c_D(\bot)}{\to} R_0\|S_1 \overset{r_A(d')}{\to} R_0\|T_{d'1}$.

So $\tau_I(\partial_H(R_0\|S_0)) \overset{r_A(d)}{\to} \overset{\tau}{\to} \overset{\tau}{\to} \overset{r_A(d')}{\to} \tau_I(\partial_H(R_0\|T_{d'1}))$.

7.1.1

$$\rho_f(\langle X \mid X=aX+bX\rangle)$$
$$\overset{\text{RDP}}{=} \rho_f(a\langle X \mid X=aX+bX\rangle + b\langle X \mid X=aX+bX\rangle)$$
$$\overset{\text{RN1,3,4}}{=} c\rho_f(\langle X \mid X=aX+bX\rangle) + c\rho_f(\langle X \mid X=aX+bX\rangle)$$
$$\overset{\text{A3}}{=} c\rho_f(\langle X \mid X=aX+bX\rangle).$$

So by RSP, $\rho_f(\langle X \mid X=aX+bX\rangle) = \langle Y \mid Y=cY\rangle$.

7.1.2 Sketch: First prove that $\rho_{g\circ f}(t) = \rho_g(\rho_f(t))$ holds for process terms s and t in BPA extended with δ and τ, by induction with respect to the size of t. Next, observe that each process term in ACP_τ with renaming is provably equal to a process term in BPA extended with δ and τ.

7.2.2

$$action(0, push) \triangleq on \qquad effect(0, push) \triangleq 1$$
$$action(1, push) \triangleq off \qquad effect(1, push) \triangleq 0$$

(The definitions of $action$ and $effect$ for the on and off are not relevant.)

$$\lambda_0(\langle X \mid X=push\cdot X\rangle) \overset{\text{RDP}}{=} \lambda_0(push \cdot \langle X \mid X=push\cdot X\rangle)$$
$$\overset{\text{SO4}}{=} on \cdot \lambda_1(\langle X \mid X=push\cdot X\rangle)$$
$$\overset{\text{RDP}}{=} on \cdot \lambda_1(push \cdot \langle X \mid X=push\cdot X\rangle)$$
$$\overset{\text{SO4}}{=} on \cdot off \cdot \lambda_0(\langle X \mid X=push\cdot X\rangle).$$

7.2.3 Let $state(s_k, c) \triangleq s_{k+1}$ for $k \in \mathbb{N}$. Moreover, let

$$action(s_0, c)\, action(s_1, c)\, action(s_2, c) \ldots$$

be a non-repetitive sequence of a's and b's. Then $\lambda_{s_0}(\langle X \mid X=cX\rangle)$ has a non-regular process graph.

7.2.4 $t = switch \cdot t + \sum_{d \in \Delta} read(d) \cdot t$ can be derived from $\mathcal{E}_{ACP}$, RDP, and commutativity of the merge. So

$$\lambda_0(t) \overset{\text{SO1-4}}{=} on \cdot \lambda_1(t) + \sum_{d \in \Delta} lost \cdot \lambda_0(t) \overset{\text{A3}}{=} lost \cdot \lambda_0(t) + on \cdot \lambda_1(t)$$
$$\lambda_1(t) \overset{\text{SO1-4}}{=} off \cdot \lambda_0(t) + \sum_{d \in \Delta} read(d) \cdot \lambda_1(t).$$

7.3.1 The weight of a transition $t \overset{a}{\to} t'$ or $t \overset{a}{\to} \checkmark$ is the number of occurrences of priority and unless operators in t.

7.3.2 Θ occurs at the right-hand side of the conclusion of the second transition rule for the priority operator. Furthermore, in the second transition rule for the priority operator, the argument x of the source $\Theta(x)$ is the left-hand side of the negative premises. This combination violates the RBB cool format.

7.3.3 $\Theta(a(b + c)) \overset{\text{TH4}}{=} \Theta(a)\Theta(b + c) \overset{\text{TH1}}{=} a\Theta(b + c) \overset{\text{TH3}}{=} a(\Theta(b) \triangleleft c + \Theta(c) \triangleleft b) \overset{\text{TH1}}{=}$
$a(b \triangleleft c + c \triangleleft b) \overset{\text{P1,2}}{=} a(\delta + c) \overset{\text{A6}}{=} ac.$

$\Theta(a(\tau(b + c) + b)) \overset{\text{TH4}}{=} \Theta(a)\Theta(\tau(b + c) + b) \overset{\text{TH1,3}}{=} a(\Theta(\tau(b+c)) \triangleleft b + \Theta(b) \triangleleft (\tau(b + c))) \overset{\text{TH1,4}}{=} a((\Theta(\tau)\Theta(b + c)) \triangleleft b + b \triangleleft (\tau(b + c))) \overset{\text{TH1,3,P8}}{=} a((\tau(\Theta(b) \triangleleft c + \Theta(c) \triangleleft b)) \triangleleft b + b \triangleleft \tau) \overset{\text{TH1,P2}}{=} a((\tau(b \triangleleft c + c \triangleleft b)) \triangleleft b + \delta) \overset{\text{A6,P1,2}}{=} a((\tau(\delta + c)) \triangleleft b) \overset{\text{A6,P6}}{=} a((\tau \triangleleft b)c) \overset{\text{P1}}{=} a(\tau c) \overset{\text{A4,B1}}{=} ac.$

7.3.4

$$\partial_{\{a,b\}}(\langle X \mid X = aX \rangle \| \langle Y \mid Y = bY \rangle)$$
$$= \partial_{\{a,b\}}(a\langle X \mid X = aX \rangle \| \langle Y \mid Y = bY \rangle + b\langle Y \mid Y = bY \rangle \| \langle X \mid X = aX \rangle$$
$$+ c\langle X \mid X = aX \rangle \| \langle Y \mid Y = bY \rangle)$$
$$= c\partial_{\{a,b\}}(\langle X \mid X = aX \rangle \| \langle Y \mid Y = bY \rangle)$$

$$\theta(\langle X \mid X = aX \rangle \| \langle Y \mid Y = bY \rangle)$$
$$= \theta(a\langle X \mid X = aX \rangle \| \langle Y \mid Y = bY \rangle + b\langle Y \mid Y = bY \rangle \| \langle X \mid X = aX \rangle$$
$$+ c\langle X \mid X = aX \rangle \| \langle Y \mid Y = bY \rangle)$$
$$= c\theta(\langle X \mid X = aX \rangle \| \langle Y \mid Y = bY \rangle).$$

So by RSP,

$$\partial_{\{a,b\}}(\langle X \mid X = aX \rangle \| \langle Y \mid Y = bY \rangle) = \langle Z \mid Z = cZ \rangle$$
$$= \theta(\langle X \mid X = aX \rangle \| \langle Y \mid Y = bY \rangle).$$

7.3.5 Let the guarded linear recursive specification E be defined by

$$X_i = a_{i1} X_{i1} + \cdots + a_{ik_i} X_{ik_i} + b_{i1} + \cdots + b_{i\ell_i}$$

for $i \in \{1, \ldots, n\}$. Let K_i and L_j consist of the indices $\alpha \in \{1, \ldots, k_i\}$ and $\beta \in \{1, \ldots, \ell_i\}$ for which $a_{i\alpha}$ and $b_{i\beta}$ are maximal in $\{a_{i1}, \ldots, a_{ik_i}, b_{i1}, \ldots, b_{i\ell_i}\}$, with respect to the partial order on atomic actions. The linear recursive specification F is defined to consist of

$$Y_i = \sum_{\alpha \in K_i} a_{i\alpha} Y_{i\alpha} + \sum_{\beta \in L_i} b_{i\beta}$$

for $i \in \{1, \ldots, n\}$. Since E is guarded, it follows that F is also guarded.

$$\theta(\langle X_i | E \rangle)$$
$$\overset{\text{RDP}}{=} \theta(a_{i1}\langle X_{i1} | E \rangle + \cdots + a_{ik_i}\langle X_{ik_i} | E \rangle + b_{i1} + \cdots + b_{i\ell_i})$$
$$\overset{\text{TH1-4,P1-8}}{=} \sum_{\alpha \in K_i} a_{i\alpha} \theta(\langle X_{i\alpha} | E \rangle) + \sum_{\beta \in L_i} b_{i\beta}.$$

Hence, replacing Y_i by $\Theta(\langle X_i|E\rangle)$ for $i \in \{1,\ldots,n\}$ is a solution for F. So by RSP, $\Theta(\langle X_1|E\rangle) = \langle Y_1|F\rangle$.

7.3.6 The following axioms originate from [35]:

$$
\begin{array}{ll}
(1) & \Theta(v) = v \\
(2) & \Theta(\delta) = \delta \\
(3) & \Theta(v \cdot x + v \cdot y + z) = \Theta(v \cdot x + z) + \Theta(v \cdot y + z) \\
(4) & \Theta(v \cdot x + v + z) = \Theta(v \cdot x + z) + \Theta(v + z) \\
(5) & \Theta(v \cdot x + w \cdot y + z) = \Theta(w \cdot y + z) \\
(6) & \Theta(v \cdot x + w + z) = \Theta(w + z) \\
(7) & \Theta(v + w \cdot y + z) = \Theta(w \cdot y + z) \\
(8) & \Theta(v + w + z) = \Theta(w + z) \\
(9) & \Theta(\sum_{i=1}^{m} v_i \cdot x_i + \sum_{j=1}^{n} w_j) = \sum_{i=1}^{m} v_i \cdot \Theta(x_i) + \sum_{j=1}^{n} w_j
\end{array}
$$

In axioms (5)-(8), $v < w$. In axiom (9), $v_1,\ldots,v_m,w_1,\ldots,w_n$ are distinct atomic actions and pairwise incomparable.

7.3.7 It suffices to prove that each process term t in ACP_τ with guarded linear recursion and the alt operator is provably equal to a process term $\langle X|E\rangle$ with E a guarded linear recursive specification. Namely, then the desired completeness result follows from the fact that if $\langle X_1|E_1\rangle \leftrightarrow_{rb} \langle Y_1|E_2\rangle$ for guarded linear recursive specifications E_1 and E_2, then $\langle X_1|E_1\rangle = \langle Y_1|E_2\rangle$ can be derived from $\mathcal{E}_{ACP} + B1, 2 + RDP, RSP$; see the proof of Theorem 5.3.2.

Apply structural induction with respect to the size of t. In comparison to the completeness proof of Theorem 5.6.2, the only new case (where the axioms ALT1-4 for alt from the solution to Exercise 3.4.11 are needed) is when $t \equiv alt(s_1, s_2)$. By induction it may be assumed that $s_1 = \langle X_1|E_1\rangle$ and $s_2 = \langle Y_1|E_2\rangle$ with E_1 and E_2 guarded linear recursive specifications, so $t = alt(\langle X_1|E_1\rangle, \langle Y_1|E_2\rangle)$. Let E_1 consist of

$$X_i = a_{i1}X_{i1} + \cdots + a_{ik_i}X_{ik_i} + b_{i1} + \cdots + b_{i\ell_i}$$

for $i \in \{1,\ldots,M\}$, and E_2 of

$$Y_j = c_{j1}Y_{j1} + \cdots + c_{jm_j}Y_{jm_j} + d_{j1} + \cdots + d_{jn_j}$$

for $j \in \{1,\ldots,N\}$, where the recursion variables X_i and Y_j are all distinct. The recursive specification F is defined to consist of E_1 and E_2 together with

$$
\begin{aligned}
V_i^j &= a_{i1}W_j^{i1} + \cdots + a_{ik_i}W_j^{ik_i} + b_{i1}Y_j + \cdots + b_{i\ell_i}Y_j \\
W_j^i &= c_{j1}V_i^{j1} + \cdots + c_{jm_j}V_i^{jm_j} + d_{j1}X_i + \cdots + d_{jn_j}X_i
\end{aligned}
$$

for $i \in \{1,\ldots,M\}$ and $j \in \{1,\ldots,N\}$. Since E_1 and E_2 are guarded, it follows that F is also guarded.

$$
\begin{aligned}
&alt(\langle X_i|E_1\rangle, \langle Y_j|E_2\rangle) \\
\overset{RDP}{=}\; &alt(a_{i1}\langle X_{i1}|E_1\rangle + \cdots + a_{ik_i}\langle X_{ik_i}|E_1\rangle + b_{i1} + \cdots + b_{i\ell_i}, \langle Y_j|E_2\rangle) \\
\overset{ALT1\text{-}4}{=}\; &a_{i1}alt(\langle Y_j|E_2\rangle, \langle X_{i1}|E_1\rangle) + \cdots + a_{ik_i}alt(\langle Y_j|E_2\rangle, \langle X_{ik_i}|E_1\rangle) \\
&+ b_{i1}\langle Y_j|E_2\rangle + \cdots + b_{i\ell_i}\langle Y_j|E_2\rangle
\end{aligned}
$$

$$
\begin{aligned}
&alt(\langle Y_j|E_2\rangle, \langle X_i|E_1\rangle) \\
\overset{RDP}{=}\; &alt(c_{j1}\langle Y_{j1}|E_2\rangle + \cdots + c_{jm_j}\langle Y_{jm_j}|E_2\rangle + d_{j1} + \cdots + d_{jn_j}, \langle X_i|E_1\rangle) \\
\overset{ALT1\text{-}4}{=}\; &c_{j1}alt(\langle X_i|E_1\rangle, \langle Y_{j1}|E_2\rangle) + \cdots + c_{jm_j}alt(\langle X_i|E_1\rangle, \langle Y_{jm_j}|E_2\rangle) \\
&+ d_{j1}\langle X_i|E_1\rangle + \cdots + d_{jn_j}\langle X_i|E_1\rangle.
\end{aligned}
$$

Hence, replacing V_i^j by $alt(\langle X_i|E_1\rangle, \langle Y_j|E_2\rangle)$, W_j^i by $alt(\langle Y_j|E_2\rangle, \langle X_i|E_1\rangle)$, X_i by $\langle X_i|E_1\rangle$, and Y_j by $\langle Y_j|E_2\rangle$ for $i \in \{1, \ldots, M\}$ and $j \in \{1, \ldots, N\}$ is a solution for F. So by RSP, $alt(\langle X_1|E_1\rangle, \langle Y_1|E_2\rangle) = \langle V_1^1|F\rangle$.

A.1.1 Some typical closed terms: a, $f(a,a)$, $g(a)$, $f(f(a,a),f(a,a))$, $f(f(a,a),g(a))$, $f(g(a),f(a,a))$, $f(g(a),g(a))$, $g(f(a,a))$, $g(g(a))$, $f(g(f(a,a)),g(g(a)))$, $\ldots$

A.1.2

- $\sigma(x) \equiv a$ and $\sigma(y) \equiv b$;
- no;
- $\sigma(x) \equiv b$, $\sigma(y) \equiv b$, and $\sigma(z) \equiv b$;
- no.

A.2.1

- $f(b,c,a) = b = f(b,c,b)$;
- $f(a,c,b) = f(c,a,b) = f(b,c,a) = b$;
- $f(c,c,f(c,c,b)) = f(c,c,f(b,c,c)) = f(c,c,b) = f(b,c,c) = b$.

A.3.1 sound, not complete; neither sound nor complete; sound and complete; complete, not sound.

A.3.2 $S(S(S(0))) + S(0) = S(S(S(S(0))) + 0) = S(S(S(S(0))))$.
$S(S(0)) \cdot S(S(0)) = (S(S(0)) \cdot S(0)) + S(S(0)) = (S(S(0)) \cdot 0 + S(S(0))) + S(S(0)) = (0 + S(S(0))) + S(S(0)) = S(0 + S(0)) + S(S(0)) = S(S(0 + 0)) + S(S(0)) = S(S(0)) + S(S(0)) = S(S(S(0)) + S(0)) = S(S(S(S(0)) + 0)) = S(S(S(S(0))))$.

A.3.3

- $\{[a], [b]\}$;
- $\{[a]\}$;
- $\{[f^k(a)] \mid k \in \mathbb{N}\}$;
- $\{[a], [f(a)]\}$;
- $\emptyset$.

A.3.4 $S([0]) + S([0]) \equiv [S(0)] + [S(0)] \equiv [S(0) + S(0)] \equiv [S(S(0) + 0)] \equiv [S(S(0))]$.

A.3.5 yes; no (e.g., $x = y$); yes; yes; no (e.g., $x = y$).

A.4.1 $S(0) + S(0) \overset{(2)}{\to} S(S(0) + 0) \overset{(1)}{\to} S(S(0))$. Now use Example A.4.1.

A.4.2 (SUBSTITUTION)

$weight(s + 0) \triangleq weight(s) + weight(0)^2 > weight(s)$;

$weight(s + S(t)) \triangleq weight(s) + weight(S(t))^2 = weight(s) + (weight(t) + 1)^2 > weight(s) + weight(t)^2 + 1 = weight(s + t) + 1 \triangleq weight(S(s + t))$;

$weight(s \cdot 0) \triangleq weight(s)^2 \cdot weight(0)^2 > weight(0)$;

$weight(s \cdot S(t)) \triangleq weight(s)^2 \cdot weight(S(t))^2 = weight(s)^2 \cdot (weight(t) + 1)^2 > weight(s)^2 \cdot weight(t)^2 + weight(s)^2 \triangleq weight((s \cdot t) + s)$.

(CONTEXT)

If $weight(t) > weight(t')$, then clearly $weight(S(t)) > weight(S(t'))$, $weight(s + t) > weight(s + t')$, $weight(t + s) > weight(t' + s)$, $weight(s \cdot t) > weight(s \cdot t')$, and $weight(t \cdot s) > weight(t' \cdot s)$.

A.4.3 Apply structural induction with respect to the size of t. If t is of the form $u + u'$ or $u \cdot u'$, then by induction t is not a normal form, so $s + t$ and $s \cdot t$ are not normal forms. So it can be assumed that $t \equiv S^k(0)$ for some $k \in \mathbb{N}$. Since $s + 0 \to s$ and $s + S^{\ell+1}(0) \to S(s + S^\ell(0))$ for $\ell \in \mathbb{N}$, $s + t$ is not a normal form. Moreover, since $s \cdot 0 \to 0$ and $s \cdot S^{\ell+1}(0) \to (s \cdot S^\ell(0)) + s$ for $\ell \in \mathbb{N}$, $s \cdot t$ is not a normal form.

A.4.4 $0 \cdot 0 \overset{(3)}{\to} 0$, while both $0 \cdot 0$ and 0 have weight 1.

A.4.5 $S(0) + S(S(0)) =_{AC} S(S(0)) + S(0) \overset{(2)}{\to} S(S(S(0)) + 0) \overset{(1)}{\to} S(S(S(0)))$.

A.4.6 Add the rewrite rule $g(h(a)) \to h(a)$.

B.1.1 $\emptyset$; $\{f^k(a)P \mid k \in \mathbb{N}\}$; $\emptyset$; $\{aP, bQ\}$.

B.2.1 $\langle\{bQ\}, \emptyset\rangle$;
$\quad\langle\emptyset, \{aP\}\rangle$;
$\quad\langle\{aP\}, \emptyset\rangle$;
$\quad\langle\emptyset, \{aP, bQ\}\rangle$, $\langle\{aP\}, \emptyset\rangle$, and $\langle\{bQ\}, \emptyset\rangle$;
$\quad\langle\emptyset, \{aP, aQ, bP, bQ\}\rangle$ and $\langle\{aP, bQ\}, \emptyset\rangle$.

B.2.2
- C_0 and C_1 are $\emptyset$, while C_α is $\{bQ\}$ for $\alpha \geq 2$; U_0 is $\{aP, aQ, bP, bQ\}$, U_1 is $\{bQ\}$, and U_α is $\emptyset$ for $\alpha \geq 2$.
  The least three-valued stable model is $\langle\{bQ\}, \emptyset\rangle$.
- C_α is $\emptyset$ for $\alpha \geq 0$; U_0 is $\{aP, aQ, bP, bQ\}$ and U_α is $\{aP\}$ for $\alpha \geq 1$.
  The least three-valued stable model is $\langle\emptyset, \{aP\}\rangle$.
- C_0 is $\emptyset$ and C_α is $\{aP\}$ for $\alpha \geq 1$; U_0 is $\{aP, aQ, bP, bQ\}$ and U_α is $\emptyset$ for $\alpha \geq 1$.
  The least three-valued stable model is $\langle\{aP\}, \emptyset\rangle$.
- C_α is $\emptyset$ for $\alpha \geq 0$; U_0 is $\{aP, aQ, bP, bQ\}$ and U_α is $\{aP, bQ\}$ for $\alpha \geq 1$.
  The least three-valued stable model is $\langle\emptyset, \{aP, bQ\}\rangle$.
- C_α is $\emptyset$ for $\alpha \geq 0$; U_α is $\{aP, aQ, bP, bQ\}$ for $\alpha \geq 0$.
  The least three-valued stable model is $\langle\emptyset, \{aP, aQ, bP, bQ\}\rangle$.

B.2.3 yes; no; yes; no; no.

B.2.4 yes; no; no; no; no.

B.2.5 The third TSS in Exercise B.2.1.

B.2.6 For $k \in \mathbb{N}$, define the weight of transitions $f^k(a)P$ to be k and the weight of transitions $f^k(a)Q$ to be $k + 1$. This constitutes a stratification.
  In the three-valued stable model for the TSS, the true transitions are $f^{2k+1}(a)P$ and $f^{2k}(a)Q$ for $k \in \mathbb{N}$; there are no unknown transitions.

B.3.1 In all three processes, let s_0 represent the root state:
- $\{s_0\}$;
- $\{s_k \overset{a}{\to} s_{k+1} \mid k \in \mathbb{N}\}$;
- $\{s_0 \overset{a}{\to} s_0\}$.

B.3.2 $s_0 \, \mathcal{B} \, s$ and $s_1 \, \mathcal{B} \, s$.

B.3.3 $f^k(a) \underset{}{\leftrightarrow} f^\ell(a)$ if and only if $k - \ell$ is even, for $k, \ell \in \mathbb{N}$. Hence, $s \underset{}{\leftrightarrow} t$ implies $f(s) \underset{}{\leftrightarrow} f(t)$.

B.3.4
- $a \underset{}{\leftrightarrow} b$ but $f(a) \underset{}{\not\leftrightarrow} f(b)$: $f(a)P$ holds while $f(b)P$ does not hold.
  The transition rule is not panth because its source contains two function symbols.
- $a \underset{}{\leftrightarrow} a$ and $a \underset{}{\leftrightarrow} b$, but $g(a, a) \underset{}{\not\leftrightarrow} g(a, b)$: $g(a, a)P$ holds while $g(a, b)P$ does not hold.
  The transition rule is not panth because its source contains two occurrences of the variable x.
- $a \underset{}{\leftrightarrow} b$ but $f(a) \underset{}{\not\leftrightarrow} f(b)$: $f(a)P$ holds while $f(b)P$ does not hold.
  The second transition rule is not panth because the variable y occurs both in the source and as the right-hand side of the premise.
- $a \underset{}{\leftrightarrow} b$ but $f(a) \underset{}{\not\leftrightarrow} f(b)$: $f(a)P$ holds while $f(b)P$ does not hold.
  The second transition rule is not panth because the right-hand side a of the premise is not a single variable.

- $a \leftrightarrow a$ and $a \leftrightarrow b$, but $g(a,a) \not\leftrightarrow g(a,b)$: $g(a,a)P$ holds while $g(a,b)P$ does not hold.
  The second transition rule is not panth because the variable y occurs as the right-hand side of both premises.

B.4.1
- yes: $s_0 \, B \, s_2$;
- yes: $s_0 \, B \, s_2$ and $s_1 \, B \, s_2$;
- no;
- yes: $s_0 \, B \, s_2$, $s_0 \, B \, s_3$, and $s_1 \, B \, s_4$.

B.4.2 It suffices to show that the two processes with root states s_1 and s_6, respectively, are branching bisimilar. This follows from the branching bisimulation relation B defined by $s_1 \, B \, s_6$, $s_2 \, B \, s_7$, $s_3 \, B \, s_6$, and $s_4 \, B \, s_7$.

B.4.3 no; no; no; yes.

B.4.4
- $f(a) \not\leftrightarrow_{rb} f(b)$ follows from the fact that $f(d)P$ holds while $f(c)P$ does not hold. f occurs at the right-hand side of the conclusion of the fifth rule. In the sixth rule, the argument of the source $f(x)$ occurs as the left-hand side of the premise. Since there is no patience rule for the argument of f, this combination violates the RBB cool format.
- $f(a) \not\leftrightarrow_{rb} f(b)$ follows from the fact that $f(c)P$ holds while $f(d)P$ does not hold. f occurs at the right-hand side of the conclusion of the fifth rule. In the sixth rule the argument of the source $f(x)$ occurs as the left-hand side of the negative premise. This combination violates the RBB cool format.
- $f(a) \not\leftrightarrow_{rb} f(b)$ follows from the fact that $f(d)P$ holds while $f(c)P$ and $f(e)P$ do not hold.
  f occurs at the right-hand side of the conclusion of the ninth rule. In the tenth rule the argument of the source $f(x)$ occurs as the left-hand side of the two premises. This combination violates the RBB cool format.

B.5.1 no (aP); yes; no (aP); yes; no $(a \xrightarrow{c} b)$; yes.

B.5.2 $-$; yes; no; yes; no; $-$.

B.5.3 no; yes; yes; yes; $-$; yes and no.

B.5.4 no; yes; no; yes; no; yes.

B.5.5 The variables x and y in the second transition rule are both source-dependent: x occurs in the source, so it is source-dependent; hence, the premise $x \xrightarrow{c} y$ ensures that y is source-dependent.
The original TSS generates $\{f^k(a) \xrightarrow{c} f^k(a) \mid k \in \mathbb{N}\}$.
The extended TSS generates $\{f^k(a) \xrightarrow{c} f^k(a) \mid k \in \mathbb{N}\} \cup \{f^k(b) \xrightarrow{c} f^k(b) \mid k \in \mathbb{N}\}$.

B.6.1 $\langle a \rangle (\langle b \rangle P \wedge \langle c \rangle P)$.

References

1. P. Abdulla, A. Annichini, and A. Bouajjani. Symbolic verification of lossy channel systems: application to the bounded retransmission protocol. In W.R. Cleaveland, ed., *Proceedings 5th Conference on Tools and Algorithms for the Construction and Analysis of Systems (TACAS'99)*, Amsterdam, LNCS 1579, pp. 208–222. Springer, 1999.
2. L. Aceto, W.J. Fokkink, R.J. van Glabbeek, and A. Ingólfsdóttir. Axiomatizing prefix iteration with silent steps. *Information and Computation*, 127(1):26–40, 1996.
3. L. Aceto, W.J. Fokkink, and C. Verhoef. Structural operational semantics. In J.A. Bergstra, A. Ponse, and S.A. Smolka, eds., *Handbook of Process Algebra*. Elsevier, 2000. To appear.
4. R. Alur, C. Courcoubetis, and D.L. Dill. Model-checking in dense real-time. *Information and Computation*, 104(1):2–34, 1993.
5. R. Alur and D.L. Dill. A theory of timed automata. *Theoretical Computer Science*, 126(2):183–235, 1994.
6. R. Alur and T.A. Henzinger, editors. *Proceedings 8th Conference on Computer-Aided Verification (CAV'96)*, New Brunswick, LNCS 1102. Springer, 1996.
7. M.A. Arbib, A.J. Kfoury, and R.N. Moll. *An Introduction to Formal Language Theory*. Springer, 1988.
8. F. Baader and T. Nipkow. *Term Rewriting and All That*. Cambridge University Press, 1998.
9. J.W. Backus. The syntax and semantics of the proposed international algebraic language of the Zurich ACM-GAMM conference. In *Proceedings ICIP*, pp. 125–131. UNESCO, 1960.
10. J.C.M. Baeten. *Procesalgebra: een Formalisme voor Parallelle, Communicerende Processen*. Kluwer, 1986. In Dutch.
11. J.C.M. Baeten, editor. *Applications of Process Algebra*. Cambridge Tracts in Theoretical Computer Science 17. Cambridge University Press, 1990.
12. J.C.M. Baeten and J.A. Bergstra. Global renaming operators in concrete process algebra. *Information and Computation*, 78(3):205–245, 1988.
13. J.C.M. Baeten and J.A. Bergstra. Process algebra with signals and conditions. In M. Broy, editor, *Proceedings Summer School on Programming and Mathematical Methods*, Marktoberdorf, NATO ASI Series F88, pp. 273–323. Springer, 1991.
14. J.C.M. Baeten and J.A. Bergstra. Real time process algebra. *Formal Aspects of Computing*, 3(2):142–188, 1991.
15. J.C.M. Baeten and J.A. Bergstra. Recursive process definitions with the state operator. *Theoretical Computer Science*, 82(2):285–302, 1991.
16. J.C.M. Baeten and J.A. Bergstra. Discrete time process algebra, *Formal Aspects of Computing*, 8(2):188–208, 1996.

17. J.C.M. Baeten and J.A. Bergstra. Process algebra with propositional signals. *Theoretical Computer Science*, 177(2):381–405, 1997.
18. J.C.M. Baeten, J.A. Bergstra, and J.W. Klop. Syntax and defining equations for an interrupt mechanism in process algebra. *Fundamenta Informaticae*, 9(2):127–167, 1986.
19. J.C.M. Baeten, J.A. Bergstra, and J.W. Klop. On the consistency of Koomen's fair abstraction rule. *Theoretical Computer Science*, 51(1/2):129–176, 1987.
20. J.C.M. Baeten, J.A. Bergstra, and J.W. Klop. Ready trace semantics for concrete process algebra with the priority operator. *The Computer Journal*, 30(6):498–506, 1987.
21. J.C.M. Baeten, J.A. Bergstra, and J.W. Klop. Conditional axioms and α/β-calculus in process algebra. In M. Wirsing, ed., *Proceedings 3rd IFIP Conference on Formal Description of Programming Concepts*, Ebberup, pp. 53–75. North-Holland, 1987.
22. J.C.M. Baeten, J.A. Bergstra, and J.W. Klop. Decidability of bisimulation equivalence for processes generating context-free languages. *Journal of the ACM*, 40(3):653–682, 1993.
23. J.C.M. Baeten, J.A. Bergstra, J.W. Klop, and W.P. Weijland. Term-rewriting systems with rule priorities. *Theoretical Computer Science*, 67(2/3):283–301, 1989.
24. J.C.M. Baeten, J.A. Bergstra, and S.A. Smolka. Axiomatizing probabilistic processes: ACP with generative probabilities. *Information and Computation*, 121(2):234–255, 1995.
25. J.C.M. Baeten and R.J. van Glabbeek. Another look at abstraction in process algebra. In T. Ottmann, ed., *Proceedings 14th Colloquium on Automata, Languages and Programming (ICALP'87)*, Karlsruhe, LNCS 267, pp. 84–94. Springer, 1987.
26. J.C.M. Baeten and C. Verhoef. A congruence theorem for structured operational semantics with predicates. In [54], pp. 477–492.
27. J.C.M. Baeten and C. Verhoef. Concrete process algebra. In S. Abramsky, D.M. Gabbay, and T.S.E. Maibaum, eds., *Handbook of Logic in Computer Science, Volume IV, Syntactical Methods*, pp. 149–268. Oxford University Press, 1995.
28. J.C.M. Baeten and W.P. Weijland. *Process Algebra*. Cambridge Tracts in Theoretical Computer Science 18. Cambridge University Press, 1990.
29. J.W. de Bakker and J.I. Zucker. Denotational semantics of concurrency. In *Proceedings 14th ACM Symposium on Theory of Computing (STOC'84)*, San Francisco, pp. 153–158. ACM, 1982.
30. J.W. de Bakker and J.I. Zucker. Processes and the denotational semantics of concurrency. *Information and Control*, 54(1/2):70–120, 1982.
31. K.A. Bartlett, R.A. Scantlebury, and P.T. Wilkinson. A note on reliable full-duplex transmission over half-duplex links. *Communications of the ACM*, 12(5):260–261, 1969.
32. T. Basten. Branching bisimilarity is an equivalence indeed! *Information Processing Letters*, 58(3):141–147, 1996.
33. H. Bekič. Towards a mathematical theory of processes. Report TR 25.125, IBM Vienna Laboratory, 1971. Also appeared in C.B. Jones, ed., *Programming Languages and their Definition: Selected Papers of H. Bekič*, LNCS 177, pp. 168–206, 1984.
34. J. Bengtsson, K.G. Larsen, F. Larsson, P. Pettersson, and Wang Yi. UPPAAL – a tool suite for automatic verification of real-time systems. In R. Alur, T.A. Henzinger, and E.D. Sontag, eds., *Proceedings 3rd Workshop on Verification and Control of Hybrid Systems*, New Brunswick, LNCS 1066, pp. 232–243. Springer, 1995.

35. J.A. Bergstra. Put and get, primitives for synchronous unreliable message passing. Logic Group Preprints 3, Utrecht University, 1985.

36. J.A. Bergstra, I. Bethke, and P.H. Rodenburg. A propositional logic with 4 values: true, false, divergent and meaningless. *Journal of Applied Non-Classical Logics*, 5(2):199–217, 1995.

37. J.A. Bergstra, W.J. Fokkink, and A. Ponse. Process algebra with recursive operations. In J.A. Bergstra, A. Ponse, and S.A. Smolka, eds., *Handbook of Process Algebra*. Elsevier, 2000. To appear.

38. J.A. Bergstra, J. Heering, and P. Klint, editors. *Algebraic Specification*. ACM Press Frontier Series. ACM/Addison Wesley, 1989.

39. J.A. Bergstra, J. Heering, and P. Klint. Module algebra. *Journal of the ACM*, 37(2):335–372, 1990.

40. J.A. Bergstra, J.A. Hillebrand, and A. Ponse. Grid protocols based on synchronous communication. *Science of Computer Programming*, 29(1/2):199–233, 1997.

41. J.A. Bergstra and J.W. Klop. Process algebra for synchronous communication. *Information and Control*, 60(1/3):109–137, 1984.

42. J.A. Bergstra and J.W. Klop. The algebra of recursively defined processes and the algebra of regular processes. In J. Paredaens, ed., *Proceedings 11th Colloquium on Automata, Languages and Programming (ICALP'84)*, Antwerp, LNCS 172, pp. 82–95. Springer, 1984.

43. J.A. Bergstra and J.W. Klop. Algebra of communicating processes with abstraction. *Theoretical Computer Science*, 37(1):77–121, 1985.

44. J.A. Bergstra and J.W. Klop. Verification of an alternating bit protocol by means of process algebra. In W. Bibel and K.P. Jantke, eds., *Proceedings Spring School on Mathematical Methods of Specification and Synthesis of Software Systems '85*, Wendisch-Rietz, LNCS 215, pp. 9–23. Springer, 1986.

45. J.A. Bergstra and J.W. Klop. A complete inference system for regular processes with silent moves. In F.R. Drake and J.K. Truss, eds., *Proceedings Logic Colloquium '86*, Hull, pp. 21–81. North-Holland, 1988.

46. J.A. Bergstra, J.W. Klop, and E.-R. Olderog. Readies and failures in the algebra of communicating processes. *SIAM Journal on Computing*, 17(6):1134–1177, 1988.

47. J.A. Bergstra, J.W. Klop, and J.V. Tucker. Algebraic tools for system construction. In E. Clarke and D. Kozen, eds., *Proceedings 4th Workshop on Logics of Programs*, Pittsburgh, LNCS 164, pp. 34–44. Springer, 1984.

48. J.A. Bergstra, J.W. Klop, and J.V. Tucker. Process algebra with asynchronous communication mechanisms. In S.D. Brookes, A.W. Roscoe, and G. Winskel, eds., *Proceedings Seminar on Semantics of Concurrency*, Pittsburgh, LNCS 197, pp. 76–95. Springer, 1985.

49. J.A. Bergstra and A. Ponse. Process algebra with five-valued logic. In C.S. Calude and M.J. Dinneen, eds., *Proceedings Discrete Mathematics and Theoretical Computer Science (DMTCS'99) and Computing: the Australasian Theory Symposium (CATS'99)*, Auckland, Australian Computer Science Communications 21(3):128–143. Springer, 1999.

50. J.A. Bergstra and J.V. Tucker. Top-down design and the algebra of communicating processes. *Science of Computer Programming*, 5(2):171–199, 1985.

51. J.A. Bergstra and J.V. Tucker. Equational specifications, complete term rewriting systems, and computable and semicomputable algebras. *Journal of the ACM*, 42(6):1194–1230, 1995.

52. G. Berry. The foundations of Esterel. In G.D. Plotkin, C.P. Stirling, and M. Tofte, eds., *Proof, Language and Interaction: Essays in Honour of Robin Milner*. MIT Press, 1999. To appear.

53. G. Berry and G. Gonthier. The Esterel synchronous programming language: design, semantics, implementation. *Science of Computer Programming*, 19(2):87–152, 1992.

54. E. Best, editor. *Proceedings 4th Conference on Concurrency Theory (CONCUR'93)*, Hildesheim, LNCS 715. Springer, 1993.

55. M.A. Bezem and J.F. Groote. Invariants in process algebra with data. In B. Jonsson and J. Parrow, eds., *Proceedings 5th Conference on Concurrency Theory (CONCUR'94)*, Uppsala, LNCS 836, pp. 401–416. Springer, 1994.

56. M.A. Bezem and J.F. Groote. A correctness proof of a one bit sliding window protocol in μCRL. *The Computer Journal*, 37(4):289–307, 1994.

57. B. Bloom. Structural operational semantics for weak bisimulations. *Theoretical Computer Science*, 146(1/2):25–68, 1995.

58. B. Bloom, S. Istrail, and A.R. Meyer. Bisimulation can't be traced, *Journal of the ACM*, 42(1):232–268, 1995.

59. R.N. Bol and J.F. Groote. The meaning of negative premises in transition system specifications. *Journal of the ACM*, 43(5):863–914, 1996.

60. T. Bolognesi and E. Brinksma. An introduction to LOTOS. *Computer Networks and ISDN Systems*, 14(1):25–59, 1987.

61. A. Bouali, A. Ressouche, V. Roy, and R. de Simone. The FC2TOOLS set. In [6], pp. 441–445.

62. R.S. Boyer and J.S. Moore. *A Computational Logic Handbook*. Academic Press, 1988.

63. P. Brémond-Grégoire, I. Lee, and R. Gerber. ACSR: an algebra of communicating shared resources with dense time and priorities. In [54], pp. 417–431.

64. S.D. Brookes, C.A.R. Hoare, and A.W. Roscoe. A theory of communicating sequential processes. *Journal of the ACM*, 31(3):560–599, 1984.

65. R.E. Bryant. Graph-based algorithms for boolean function manipulation. *IEEE Transactions on Computers*, 35(8):677–691, 1986.

66. C.C. Chang and H.J. Keisler. *Model Theory*. Studies in Logic and the Foundations of Mathematics 73. North-Holland, 1990.

67. S. Christensen, H. Hüttel, and C.P. Stirling. Bisimulation equivalence is decidable for all context-free processes. *Information and Computation*, 121(2):143–148, 1995.

68. A. Cimatti, E.M. Clarke, F. Giunchiglia, and M. Roveri. NuSMV: a new symbolic model verifier. In N. Halbwachs and D. Peled, eds., *Proceeding 11th Conference on Computer-Aided Verification (CAV'99)*, Trento, LNCS 1633, pp. 495–499. Springer, 1999.

69. D. Clarke, H. Ben-Abdallah, I. Lee, H.-l. Xie, and O. Sokolsky. XVERSA: an integrated graphical and textual toolset for the specification and analysis of resource-bound real-time systems. In [6], pp. 402–405.

70. E.M. Clarke and E.A. Emerson. Design and synthesis of synchronization skeletons using branching time temporal logic. In D. Kozen, ed., *Proceedings 3rd Workshop on Logics of Programs*, Yorktown Heights, LNCS 131, pp. 52–71. Springer, 1981.

71. E.M. Clarke, E.A. Emerson, and A.P. Sistla. Automatic verification of finite-state concurrent systems using temporal logic specifications. *ACM Transactions on Programming Languages and Systems*, 8(2):244–263, 1986.

72. E.M. Clarke, O. Grumberg, and D. Long. Verification tools for finite-state concurrent systems. In J.W. de Bakker, W.-P. de Roever, and G. Rozenberg, eds., *Proceedings REX School/Symposium "A Decade of Concurrency - Reflections and Perspectives"*, Noordwijkerhout, LNCS 803, pp. 124–175. Springer, 1994.

73. W.R. Cleaveland, P.M. Lewis, S.A. Smolka, and O. Sokolsky. The Concurrency Factory: a development environment for concurrent systems. In [6], pp. 398–401.

74. W.R. Cleaveland, J.G. Parrow, and B. Steffen. The Concurrency Workbench: a semantics-based tool for the verification of concurrent systems. *ACM Transactions on Programming Languages and Systems*, 15(1):36–72, 1993.

75. W.R. Cleaveland and S. Sims. The NCSU Concurrency Workbench. In [6], pp. 394–397.

76. Concurrent Systems.
http://www.comlab.ox.ac.uk/archive/concurrent.html.

77. P.R. D'Argenio, J.-P. Katoen, and E. Brinksma. An algebraic approach to the specification of stochastic systems. In D. Gries and W.-P. de Roever, eds., *Proceedings 4th IFIP Working Conference on Programming Concepts, Methods and Calculi (PROCOMET'98)*, Shelter Island, pp. 126–147. Chapman & Hall, 1998.

78. P.R. D'Argenio, J.-P. Katoen, T.C. Ruys, and J. Tretmans. The bounded retransmission protocol must be on time! In E. Brinksma, ed., *Proceedings 3rd Workshop on Tools and Algorithms for the Construction and Analysis of Systems (TACAS'97)*, Enschede, LNCS 1217, pp. 416–431. Springer, 1997.

79. C. Daws, A. Olivero, and S. Yovine. Verifying ET-LOTOS programs with KRONOS. In D. Hogrefe and S. Leue, eds., *Proceedings 7th IFIP Conference on Formal Description Techniques (FORTE'94)*, Bern, pp. 227–242. Chapman & Hall, 1994.

80. R. De Nicola and M.C.B. Hennessy. Testing equivalences for processes. *Theoretical Computer Science*, 34(1/2):83–133, 1984.

81. R. De Nicola and F.W. Vaandrager. Action versus state based logics for transition systems. In I. Guessarian, ed., *Proceedings Spring School on Semantics of Systems of Concurrent Processes*, La Roche Posay, LNCS 469, pp. 407–419. Springer, 1990.

82. R. De Nicola and F.W. Vaandrager. Three logics for branching bisimulation. *Journal of the ACM*, 42(2):458–487, 1995.

83. D.L. Dill. The Murφ verification system. In [6], pp. 390–393.

84. E.A. Emerson. Automated temporal reasoning about reactive systems. In F. Moller and G. Birtwistle, eds., *Logics for Concurrency: Structure versus Automata*, LNCS 1043, pp. 41–101. Springer, 1996.

85. E.A. Emerson and C.-L. Lei. Modalities for model checking: branching time logic strikes back. *Science of Computer Programming*, 8(3):275–306, 1987.

86. E.A. Emerson and J.Y. Halpern. "Sometimes" and "not never" revisited: on branching versus linear time. *Journal of the ACM*, 33(1):151–178, 1986.

87. E.A. Emerson and A.P. Sistla. Deciding full branching time logic. *Information and Control*, 61(3):175–201, 1984.

88. J.-C. Fernandez, H. Garavel, L. Mounier, A. Rasse, C. Rodriguez, and J. Sifakis. A toolbox for the verification of LOTOS programs. In L.A. Clarke, ed., *Proceedings 14th Conference on Software Engineering (ICSE'92)*, Melbourne, pp. 246–259. ACM, 1992.

89. W.J. Fokkink. An axiomatization for regular processes in timed branching bisimulation. *Fundamenta Informaticae*, 32(3/4):329–340, 1998.

90. W.J. Fokkink. Rooted branching bisimulation as a congruence. *Journal of Computer and System Sciences*. To appear.

91. W.J. Fokkink and R.J. van Glabbeek. Ntyft/ntyxt rules reduce to ntree rules. *Information and Computation*, 126(1):1–10, 1996.

92. W.J. Fokkink and A.S. Klusener. An effective axiomatization for real time ACP. *Information and Computation*, 122(2):286–299, 1995

93. W.J. Fokkink and C. Verhoef. A conservative look at operational semantics with variable binding. *Information and Computation*, 146(1):24–54, 1998.

94. W.J. Fokkink and H. Zantema. Basic process algebra with iteration: completeness of its equational axioms. *The Computer Journal*, 37(4):259–267, 1994.
95. Formal Methods.
    http://www.comlab.ox.ac.uk/archive/formal-methods.html.
96. L.-å. Fredlund, J.F. Groote, and H.P. Korver. Formal verification of a leader election protocol in process algebra. *Theoretical Computer Science*, 177(2):459–486, 1997.
97. A. van Gelder, K. Ross, and J.S. Schlipf. The well-founded semantics for general logic programs, *Journal of the ACM*, 38(3):620–650, 1991.
98. R.J. van Glabbeek. Bounded nondeterminism and the approximation induction principle in process algebra. In F.J. Brandeburg, G. Vidal-Naquet, and M. Wirsing, eds., *Proceedings 4th Symposium on Theoretical Aspects of Computer Science (STACS'87)*, Passau, LNCS 247, pp. 336–347. Springer, 1987.
99. R.J. van Glabbeek. The linear time – branching time spectrum. In J.C.M. Baeten and J.W. Klop, eds., *Proceedings 1st Conference on Concurrency Theory (CONCUR'90)*, Amsterdam, LNCS 458, pp. 278–297. Springer, 1990.
100. R.J. van Glabbeek. The linear time – branching time spectrum II: the semantics of sequential systems with silent moves. In [54], pp. 66–81.
101. R.J. van Glabbeek. A complete axiomatization for branching bisimulation congruence of finite-state behaviours. In A.M. Borzyszkowski and S. Sokołowski, eds., *Proceedings 18th Symposium on Mathematical Foundations of Computer Science (MFCS'93)*, Gdansk, LNCS 711, pp. 473–484. Springer, 1993.
102. R.J. van Glabbeek. What is branching time and why to use it? In M. Nielsen, ed., *The Concurrency Column, Bulletin of the EATCS*, 53:190–198, 1994.
103. R.J. van Glabbeek. The meaning of negative premises in transition system specifications II. In F. Meyer auf der Heide and B. Monien, eds., *Proceedings 23rd Colloquium on Automata, Languages and Programming (ICALP'96)*, Paderborn, LNCS 1099, pp. 502–513. Springer, 1996.
104. R.J. van Glabbeek. Personal communication. November 1997.
105. R.J. van Glabbeek and W.P. Weijland. Branching time and abstraction in bisimulation semantics. *Journal of the ACM*, 43(3):555–600, 1996.
106. M.J.C. Gordon and T.F. Melham, editors. *Introduction to HOL: a Theorem Proving Environment for Higher Order Logic*. Cambridge University Press, 1993.
107. J.F. Groote. *Process Algebra and Structured Operational Semantics*. PhD thesis, University of Amsterdam, 1991.
108. J.F. Groote. Transition system specifications with negative premises. *Theoretical Computer Science*, 118(2):263–299, 1993.
109. J.F. Groote and H.P. Korver. Correctness proof of the bakery protocol in μCRL. In *Proceedings 1st Workshop on the Algebra of Communicating Processes (ACP'94)*, Utrecht, Workshops in Computing, pp. 63–86. Springer, 1995.
110. J.F. Groote and J.C. van de Pol. A bounded retransmission protocol for large data packets: a case study in computer checked verification. In M. Wirsing and M. Nivat, eds., *Proceedings 5th Conference on Algebraic Methodology and Software Technology (AMAST'96)*, Munich, LNCS 1101, pp. 536–550. Springer, 1996.
111. J.F. Groote and A. Ponse. Process algebra with guards: combining Hoare logic with process algebra. *Formal Aspects of Computing*, 6(2):115–164, 1994.
112. J.F. Groote and A. Ponse. Syntax and semantics of μCRL. In *Proceedings 1st Workshop on the Algebra of Communicating Processes (ACP'94)*, Utrecht, Workshops in Computing, pp. 26–62. Springer, 1995.
113. J.F. Groote and M.P.A. Sellink. Confluence for process verification. *Theoretical Computer Science*, 170(1/2):47–81, 1996.

114. J.F. Groote and J. Springintveld. Focus points and convergent process operators: a proof strategy for protocol verification. In A. Arnold, ed., *Proceedings 2nd AMAST Workshop on Real-Time Systems (ARTS'95)*, Bordeaux. To appear.

115. J.F. Groote and F.W. Vaandrager. An efficient algorithm for branching bisimulation and stuttering equivalence. In M.S. Paterson, ed., *Proceedings 17th Colloquium on Automata, Languages and Programming (ICALP'90)*, Warwick, LNCS 443, pp. 626–638. Springer, 1990.

116. J.F. Groote and F.W. Vaandrager. Structured operational semantics and bisimulation as a congruence. *Information and Computation*, 100(2):202–260, 1992.

117. R. Guillemot, M. Haj-Hussein, and L. Logrippo. Executing large LOTOS specifications. In S. Aggarwal and K.K. Sabnani, *Proceedings 8th IFIP Symposium on Protocol Specification, Testing and Verification (PSTV'98)*, Atlantic City, pp. 399–410. North-Holland, 1988.

118. R.H. Hardin, Z. Har'El, and R.P. Kurshan. COSPAN. In [6], pp. 423–427.

119. L. Helmink, M.P.A. Sellink, and F.W. Vaandrager. Proof-checking a data link protocol. In H.P. Barendregt and T. Nipkow, eds., *Selected Papers 1st Workshop on Types for Proofs and Programs (TYPES'93)*, Nijmegen, LNCS 806, pp. 127–165. Springer, 1994

120. M.C.B. Hennessy. *Algebraic Theory of Processes*. MIT Press, 1988.

121. M.C.B. Hennessy and R. Milner. Algebraic laws for nondeterminism and concurrency. *Journal of the ACM*, 32(1):137–161, 1985.

122. M.C.B. Hennessy and C.P. Stirling. The power of the future perfect in program logics. *Information and Control*, 67(1/3):23–52, 1985.

123. Y. Hirshfeld and M. Jerrum. Bisimulation equivalence is decidable for normed process algebra. In J. Wiedermann, P. van Emde Boas, and M. Nielsen, eds., *Proceedings 26th Colloquium on Automata, Languages and Programming (ICALP'99)*, Prague, LNCS 1644, pp. 412–421. Springer, 1999.

124. Y. Hirshfeld, M. Jerrum, and F. Moller. A polynomial-time algorithm for deciding bisimulation equivalence of normed basic parallel processes. *Mathematical Structures in Computer Science*, 6(3):251–259, 1996.

125. C.A.R. Hoare. An axiomatic basis for computer programming. *Communications of the ACM*, 12(10):576–580,583, 1969.

126. C.A.R. Hoare. Communicating sequential processes. *Communications of the ACM*, 21(8):666–677, 1978.

127. C.A.R. Hoare. *Communicating Sequential Processes*. Prentice Hall, 1985.

128. G.J. Holzmann. *Design and Validation of Computer Protocols*. Prentice Hall, 1990.

129. P.-A. Hsiung and F. Wang. A state graph manipulator tool for real-time system specification and verification. In *Proceedings 5th Conference on Real-Time Computing Systems and Applications (RTCSA'98)*, Hiroshima, pp. 181–188. IEEE Computer Society Press, 1998.

130. G.P. Huet. Confluent reductions: abstract properties and applications to term rewriting systems. *Journal of the ACM*, 27(4):797–821, 1980.

131. G.E. Hughes and M.J. Cresswell. *A Companion to Modal Logic*. Methuen, 1984.

132. P.C. Kanellakis and S.A. Smolka. CCS expressions, finite state processes, and three problems of equivalence. *Information and Computation*, 86(1):43–68, 1990.

133. A.S. Klusener. Abstraction in real time process algebra. In J.W. de Bakker, C. Huizing, W.-P. de Roever, and G. Rozenberg, eds., *Proceedings REX Workshop "Real-Time: Theory in Practice"*, Mook, LNCS 600, pp. 325–352. Springer, 1991.

134. A.S. Klusener. The silent step in time. In W.R. Cleaveland, ed., *Proceedings 3rd Conference on Concurrency Theory (CONCUR'92)*, Stony Brook, LNCS 630, pp. 421–435. Springer, 1992.

135. D.E. Knuth and P.B. Bendix. Simple word problems in universal algebras. In J. Leech, ed., *Computational Problems in Abstract Algebra*, pp. 263–297. Pergamon Press, 1970.

136. C.J. Koomen. *A Structure Theory for Communication Network Control*. PhD thesis, Delft Technical University, 1982.

137. H.P. Korver and M.P.A. Sellink. A formal axiomatization for alphabet reasoning with parametrized processes. *Formal Aspects of Computing*, 10(1):30–42, 1998.

138. H.P. Korver and J. Springintveld. A computer-checked verification of Milner's scheduler. In M. Hagiya and J.C. Mitchell, eds., *Proceedings 2nd Symposium on Theoretical Aspects of Computer Software (TACS'94)*, Sendai, LNCS 789, pp. 161–178. Springer, 1994.

139. D. Kozen. Results on the propositional μ-calculus. *Theoretical Computer Science*, 27(3):333–354, 1983.

140. K.G. Larsen and R. Milner. A compositional protocol verification using relativized bisimulation. *Information and Computation*, 99(1):80–108, 1992.

141. K.G. Larsen and A. Skou. Bisimulation through probabilistic testing. *Information and Computation*, 94(1):1–28, 1991.

142. O. Lichtenstein and A. Pnueli. Checking that finite state concurrent programs satisfy their linear specification. In *Conference Record 12th ACM Symposium on Principles of Programming Languages (POPL'85)*, New Orleans, pp. 97–107. ACM, 1985.

143. J. Loeckx, H.-D. Ehrich, and M. Wolf. *Specification of Abstract Data Types*. Wiley/Teubner, 1996.

144. Z. Manna, N. Björner, A. Browne, E. Chang, M. Colón, L. de Alfaro, H. Devarajan, A. Kapur, J. Lee, H.B. Sipma, and T.E. Uribe. STeP: the Stanford Temporal Prover. In P.D. Mosses, M. Nielsen, and M.I. Schwartzbach, eds., *Proceedings 6th Conference on Theory and Practice of Software Development (TAPSOFT'95)*, Aarhus, LNCS 915, pp. 793–794.

145. S. Mauw and G.J. Veltink. A process specification formalism. *Fundamenta Informaticae*, 13(2):85–139, 1990.

146. S. Mauw and G.J. Veltink, editors. *Algebraic Specification of Communication Protocols*. Cambridge Tracts in Theoretical Computer Science 36. Cambridge University Press, 1993.

147. K.L. McMillan. *Symbolic Model Checking: an Approach to the State Explosion Problem*. PhD thesis, Carnegie Mellon University, 1992.

148. G.J. Milne. CIRCAL and the representation of communication, concurrency, and time. *ACM Transactions on Programming Languages and Systems*, 7(2):270–298, 1985.

149. R. Milner. Processes: a mathematical model of computing agents. In H.E. Rose and J.C. Shepherdson, eds., *Proceedings Logic Colloquium '73*, Bristol, Studies in Logic and the Foundations of Mathematics 80, pp. 157–173. North-Holland, 1975.

150. R. Milner. Synthesis of communicating behaviour. In J. Winkowski, ed., *Proceedings 7th Symposium on Mathematical Foundations of Computer Science (MFCS'78)*, Zakopane, LNCS 64, pp. 71–83. Springer, 1978.

151. R. Milner. *A Calculus of Communicating Systems*. LNCS 92, Springer, 1980.

152. R. Milner. A modal characterisation of observable machine-behaviour. In E. Astesiano and C. Böhm, eds., *Proceedings 6th Colloquium on Trees in Algebra and Programming (CAAP'81)*, Genoa, LNCS 112, pp. 25–34. Springer, 1981.

153. R. Milner. Calculi for synchrony and asynchrony. *Theoretical Computer Science*, 25(3):267–310, 1983.

154. R. Milner. A complete inference system for a class of regular behaviours. *Journal of Computer and System Sciences*, 28(3):439–466, 1984.

155. R. Milner. *Communication and Concurrency*. Prentice Hall, 1989.

156. R. Milner. A complete axiomatisation for observational congruence of finite-state behaviors. *Information and Computation*, 81(2):227–247, 1989.

157. R. Milner. *Communicating and Mobile Systems: the π-Calculus*. Cambridge University Press, 1999.

158. R. Milner, J.G. Parrow, and D. Walker. A calculus of mobile processes, part I + II. *Information and Computation*, 100(1):1–77, 1992.

159. F. Moller. The importance of the left merge operator in process algebras. In M.S. Paterson, ed., *Proceedings 17th Colloquium on Automata, Languages and Programming (ICALP'90)*, Warwick, LNCS 443, pp. 752–764. Springer, 1990.

160. F. Moller and C.M.N. Tofts. A temporal calculus of communicating systems. In J.C.M. Baeten and J.W. Klop, eds., *Proceedings 1st Conference on Concurrency Theory (CONCUR'90)*, Amsterdam, LNCS 458, pp. 401–415. Springer, 1990.

161. M.H.A. Newman. On theories with a combinatorial definition of "equivalence". *Annals of Mathematics (Series 2)*, 43(2):223–243, 1942.

162. E.-R. Olderog and C.A.R. Hoare. Specification-oriented semantics for communicating processes. *Acta Informatica*, 23(1):9–66, 1986.

163. S. Owre, J.M. Rushby, and N. Shankar. PVS: a Prototype Verification System. In D. Kapur, ed., *Proceedings 11th Conference on Automated Deduction (CADE'92)*, Saratoga Springs, LNCS 607, pp. 748–752. Springer, 1992

164. R. Paige and R.E. Tarjan. Three partition refinement algorithms. *SIAM Journal on Computing*, 16(6):973–989, 1987.

165. C.H. Papadimitriou. *Computational Complexity*. Addison-Wesley, 1994.

166. D.M.R. Park. Concurrency and automata on infinite sequences. In P. Deussen, ed., *Proceedings 5th GI (Gesellschaft für Informatik) Conference*, Karlsruhe, LNCS 104, pp. 167–183. Springer, 1981.

167. L.C. Paulson. Isabelle: the next seven hundred theorem provers. In E. Lusk and R. Overbeek, eds., *Proceedings 9th Conference on Automated Deduction (CADE'88)*, Argonne, LNCS 310, pp. 772–773. Springer, 1988.

168. C.A. Petri. *Kommunikation mit Automaten*. PhD thesis, Institut für instrumentelle Mathematik, Bonn, 1962. In German.

169. I.C.C. Phillips. Refusal testing. *Theoretical Computer Science*, 50(3):241–284, 1987.

170. D.A. Plaisted. Equational reasoning and term rewriting systems. In D. Gabbay and J. Siekmann, eds., *Handbook of Logic in Artificial Intelligence and Logic Programming*, Volume 1, pp. 273–364. Oxford University Press, 1993.

171. G.D. Plotkin. A structural approach to operational semantics. Report DAIMI FN-19, Aarhus University, 1981.

172. A. Pnueli. The temporal logic of programs. In *Proceedings 18th IEEE Symposium on Foundations of Computer Science (FOCS'77)*, Providence, pp. 46–57. IEEE Computer Society Press, 1977.

173. A. Pnueli. Linear and branching structures in the semantics and logics of reactive systems. In W. Brauer, ed., *Proceedings 12th Colloquium on Automata, Languages and Programming (ICALP'85)*, Nafplion, LNCS 194, pp. 15–32. Springer, 1985.

174. T.C. Przymusinski. On the declarative semantics of deductive databases and logic programs. In J. Minker, ed., *Foundations of Deductive Databases and Logic Programming*, Los Altos, pp. 193–216. Morgan Kaufmann, 1988.

175. T.C. Przymusinski. The well-founded semantics coincides with the three-valued stable semantics. *Fundamenta Informaticae*, 13(4):445–463, 1990.

176. Y.S. Ramakrishna, C.R. Ramakrishnan, I.V. Ramakrishnan, S.A. Smolka, T. Swift, and D.S. Warren. Efficient model checking using tabled resolution. In O. Grumberg, ed., *Proceedings 9th Conference on Computer Aided Verification (CAV'97)*, Haifa, LNCS 1254, pp. 143–154. Springer, 1997.

177. A.W. Roscoe. *The Theory and Practice of Concurrency*. Prentice Hall, 1998.

178. A. Salomaa. *Theory of Automata*. International Series of Monographs in Pure and Applied Mathematics 100. Pergamon Press, 1969.

179. D. Sangiorgi. πI: a symmetric calculus based on internal mobility. In P.D. Mosses, M. Nielsen, and M.I. Schwartzbach, eds., *Proceedings 6th Conference on Theory and Practice of Software Development (TAPSOFT'95)*, Aarhus, LNCS 915, pp. 172–186. Springer, 1995.

180. C. Shankland and M.B. van der Zwaag. The tree identify protocol of IEEE 1394 in μCRL. *Formal Aspects of Computing*, 10(6):509–531, 1998.

181. A.P. Sistla and E.M. Clarke. The complexity of propositional linear temporal logics. *Journal of the ACM*, 32(3):733–749, 1985.

182. C.P. Stirling. Modal and temporal logics for processes. In F. Moller and G. Birtwistle, eds., *Logics for Concurrency: Structure versus Automata*, LNCS 1043, pp. 149–237. Springer, 1996.

183. A.S. Tanenbaum. *Computer Networks*. Prentice Hall, 1981.

184. A. Tarski. A lattice-theoretical fixpoint theorem and its applications. *Pacific Journal of Mathematics*, 5:285–309, 1955.

185. B.C. Thompson and J.V. Tucker. Equational specification of synchronous concurrent algorithms and architectures. Report CSR 9-91, University of Wales Swansea, 1991. (Second edition, 1994).

186. C.M.N. Tofts. Describing social insect behaviour using process algebra. *Transactions of the Society for Computer Simulation*, pp. 227–283, 1992.

187. C.M.N. Tofts. Processes with probabilities, priority and time. *Formal Aspects of Computing*, 6(5):536–564, 1994.

188. A.M. Turing. On computable numbers, with an application to the Entscheidungsproblem. *Proceedings of the London Mathematical Society (Series 2)*, 42:230–265, 1936.

189. F.W. Vaandrager. Verification of two communication protocols by means of process algebra. Report CS-R8608, CWI, Amsterdam, 1986.

190. F.W. Vaandrager. *Algebraic Techniques for Concurrency and their Application*. PhD thesis, University of Amsterdam, 1990.

191. C. Verhoef. A general conservative extension theorem in process algebra, In E.-R. Olderog, ed., *Proceedings 3rd IFIP Working Conference on Programming Concepts, Methods and Calculi (PROCOMET'94)*, San Miniato, pp. 149–168. North-Holland/Elsevier, 1994.

192. C. Verhoef. A congruence theorem for structured operational semantics with predicates and negative premises. *Nordic Journal of Computing*, 2(2):274–302, 1995.

193. J.J. van Wamel. Process algebra with language matching. *Theoretical Computer Science*, 177(2):425–458, 1997.

194. M.B. van der Zwaag. Some verifications in process algebra with iota. In J.F. Groote, S.P. Luttik, and J.J. van Wamel, eds., *Proceedings 3rd Workshop on Formal Methods for Industrial Critical Systems (FMICS'98)*, Amsterdam, pp. 347–368. Stichting Mathematisch Centrum, 1998.

Index

Monographs in Theoretical Computer Science · An EATCS Series

C. Calude
Information and Randomness
An Algorithmic Perspective

K. Jensen
Coloured Petri Nets
Basic Concepts, Analysis Methods
and Practical Use, Vol. 1
2nd ed.

K. Jensen
Coloured Petri Nets
Basic Concepts, *Analysis Methods*
and Practical Use, Vol. 2

K. Jensen
Coloured Petri Nets
Basic Concepts, Analysis Methods
and *Practical Use*, Vol. 3

A. Nait Abdallah
The Logic of Partial Information

Z. Fülöp, H. Vogler
Syntax-Directed Semantics
Formal Models
Based on Tree Transducers

A. de Luca, S. Varricchio
**Finiteness and Regularity
in Semigroups
and Formal Languages**

Texts in Theoretical Computer Science · An EATCS Series

J. L. Balcázar, J. Díaz, J. Gabarró
Structural Complexity I
2nd ed. (see also overleaf, Vol. 22)

M. Garzon
Models of Massive Parallelism
Analysis of Cellular Automata
and Neural Networks

J. Hromkovič
**Communication Complexity
and Parallel Computing**

A. Leitsch
The Resolution Calculus

G. Păun, G. Rozenberg, A. Salomaa
DNA Computing
New Computing Paradigms

A. Salomaa
Public-Key Cryptography
2nd ed.

K. Sikkel
Parsing Schemata
A Framework for Specification
and Analysis of Parsing Algorithms

H. Vollmer
Introduction to Circuit Complexity
A Uniform Approach

W. Fokkink
Introduction to Process Algebra

Former volumes appeared as
EATCS Monographs on Theoretical Computer Science

Vol. 5: W. Kuich, A. Salomaa
Semirings, Automata, Languages

Vol. 6: H. Ehrig, B. Mahr
Fundamentals of Algebraic Specification 1
Equations and Initial Semantics

Vol. 7: F. Gécseg
Products of Automata

Vol. 8: F. Kröger
Temporal Logic of Programs